"An interesting insight into the world of maths anxiety and of the
Professor Nira Chamber
Th

"The publication of this admirable book is timely. It is an accessible, comprehensive and balanced introduction to the concept of maths anxiety and its impact on learners. It gives a multi-faceted perspective with sound and cautious advice. Its comprehensive collection of references would make it an essential for any researcher into maths anxiety. I found this book to be classroom friendly in that it takes a realistic view of the implications and impact of maths anxiety. Realistic and wise. Well done Heidi and Tom."

Dr Steve Chinn, Founder/Director of Maths Explained

"Heidi Kirkland and Thomas Hunt in compiling this book have provided a huge service to everyone concerned with maths anxiety, including teachers, parents, researchers, educationists, and indeed those who suffer from it and want to better understand it. They sensitively discuss, with the benefit of their considerable knowledge and experience, the nature and prevalence of the problem, how it relates to other aspects of maths learning and teaching, and importantly what can be done to improve the situation and reduce anxiety levels. They draw on a comprehensive and rigorous survey of existing research, yet the book remains very readable; quotations sprinkled throughout are especially effective in communicating the reality and variety of people's experiences of maths anxiety and how it has affected their lives. How wonderful it would be if everyone involved in teaching and learning maths could read it and act on it!"

Professor Margaret Brown OBE, Emeritus Professor at King's College London

"This is an accessible, yet thorough exploration of mathematics anxiety that distils the research for busy teachers whilst maintaining criticality. For teachers who are interested in creating inclusive classrooms, promoting positive dispositions towards mathematics and understanding their pupils better in relation to their feelings about this vitally important subject, this is a useful read, with clear practical applications."

Rhiannon Jones, Teach First

MATHS ANXIETY

Maths Anxiety is a very common experience, reported by people around the world. This book will help all teachers of Maths, from primary education to secondary and beyond, to better understand Maths Anxiety and learn how we can address this within our classrooms, creating a more positive learning environment for all.

Supported by research, case studies, and anecdotes, this highly practical book explores and provides answers to the following questions:

- What is Maths Anxiety?
- Who experiences it?
- Can I measure it?
- What is involved in Maths Anxiety?
- Who and what can influence it?
- What can we do to provide support?
- Can teachers experience Maths Anxiety too?

An essential resource for anybody teaching or supporting the learning of Maths, this book provides tried and tested interventions to apply within the classroom, alongside strategies to use at school or at home to manage and alleviate student experiences of Maths Anxiety.

Heidi Kirkland is an experienced primary teacher, Maths coordinator, and associate lecturer. She has authored numerous blogs and articles on Maths Anxiety.

Thomas Hunt is an associate professor in Psychology at the University of Derby where he leads the Mathematics Anxiety Research Group. He has authored several academic research papers and book chapters on the topic of Maths Anxiety.

MATHS ANXIETY

Solving the Equation

Heidi Kirkland and Thomas Hunt

LONDON AND NEW YORK

Designed cover image: © Getty Images

First published 2025
by Routledge
4 Park Square, Milton Park, Abingdon, Oxon OX14 4RN

and by Routledge
605 Third Avenue, New York, NY 10158

Routledge is an imprint of the Taylor & Francis Group, an informa business

© 2025 Heidi Kirkland and Thomas Hunt

The right of Heidi Kirkland and Thomas Hunt to be identified as authors of this work has been asserted in accordance with sections 77 and 78 of the Copyright, Designs and Patents Act 1988.

All rights reserved. No part of this book may be reprinted or reproduced or utilised in any form or by any electronic, mechanical, or other means, now known or hereafter invented, including photocopying and recording, or in any information storage or retrieval system, without permission in writing from the publishers.

Trademark notice: Product or corporate names may be trademarks or registered trademarks, and are used only for identification and explanation without intent to infringe.

British Library Cataloguing-in-Publication Data
A catalogue record for this book is available from the British Library

ISBN: 978-1-032-73080-6 (hbk)
ISBN: 978-1-032-73544-3 (pbk)
ISBN: 978-1-003-42654-7 (ebk)

DOI: [10.4324/9781003426547]

Typeset in Interstate
by codeMantra

In gratitude and remembrance to Dame Shirley Conran, whose legacy of unwavering compassion and tireless support marked a positive step in the fight against Maths Anxiety.

CONTENTS

	Introduction	1
1	What Is It?	7
2	Who Has It?	19
3	Can I Measure It?	35
4	What's Involved?	58
5	Who and What Can Influence It?	93
6	What Can We Do?	140
7	Can Teachers Be Maths Anxious, Too?	189
8	Can We Solve the Equation?	200
	Index	205

Introduction

As educators, we have a lot to contend with. Whether you are teaching children in the Early Years, or post-graduates, if you have been teaching for 23 years or you are a newly qualified teacher, there is so much to think about. Lesson plans, accountability, observations, attendance, parent consultations, meetings – the list is endless before we even mention students. Their attainment, their progress, their well-being, their academic needs, their lives outside of school.

Teachers and other professionals involved in Maths education don't always have the time to take a step back to understand what might be affecting students' engagement with, and enjoyment of, Maths. Maths remains a core subject and is often associated with all sorts of emotions and challenges when it comes to learning, testing, and application. One of these challenges is the experience of Maths Anxiety, which may rise and fall, bubble away under the surface, or be present without it ever really being tackled head-on.

If you have bought this book, you may be interested to know more about Maths Anxiety, or you are frantically trying to find a cure for it. Between us, we have spent 30 years or so learning about, and researching, Maths Anxiety and have pieced much of it together for you in this book. After several decades of maths anxiety research, people's general awareness and understanding of Maths Anxiety have increased, yet there is still much work to do in this regard. As you will see, Maths Anxiety is more complex than it might first appear.

While there are many views about what Maths Anxiety actually is, and therefore some disagreements among researchers and theorists, it is broadly accepted that Maths Anxiety is a negative response to a person's previous, current, or anticipated experience of Maths. This might include learning, evaluative, and applied contexts and typically comprises emotional, cognitive, and behavioural components.

Someone who is maths anxious may engage in different behaviours to usual when using or learning Maths, such as becoming avoidant, or their emotions can change and they may begin to feel worried, distressed, or helpless. They may have negative attitudes towards Maths, hold negative self-beliefs, or have a lack of motivation. They may show a change in cognition, or they may physically show signs of being anxious, such as having a dry mouth or sweaty palms. A maths anxious individual may show none, some, all, or even more of the above signs. As such, it can be difficult to define the exact nature of Maths Anxiety, but there

DOI: 10.4324/9781003426547-1

are common threads that we will explore to help you gain a clearer picture of the experience of Maths Anxiety.

Whether you are looking for practical advice or theoretical insights into Maths Anxiety (or both!) we believe this book provides just that. Starting with an in-depth look at what Maths Anxiety is, and who might experience it, we then explore how it can be measured, what is often involved in the experience, and what can influence it. By using real-life examples throughout, we aim to help you better understand the experience of Maths Anxiety and recognise it within the classroom. We then look at what we can do to help students who are experiencing Maths Anxiety, to assist you in tackling Maths Anxiety and boost positivity within your classroom. Importantly, we also acknowledge and explore how teachers can be maths anxious too and reflect on what can be done to support them.

This book will also identify what our next steps may be as educators and researchers in the fight against Maths Anxiety, and how we can increase awareness of the experience to ensure that it becomes better known, and better understood.

Chapter 1 – "What Is It?"

- Maths Anxiety as a construct
- Definitions of Maths Anxiety
- Perspectives of Maths Anxiety
- Maths Anxiety compared to other anxieties

Chapter 2 – "Who Has It?"

- Prevalence of Maths Anxiety
- Trends in prevalence
- Factors affecting prevalence

Chapter 3 – "Can I Measure It?"

- Physiological measures
- Brain activity
- Adult and adolescent questionnaires
- Child questionnaires
- Qualitative approaches

Chapter 4 – "What's Involved?"

- Attainment
- Working memory
- Attitudes
- Self-belief

- Motivation
- Avoidance

Chapter 5 – "Who And What Can Influence It?"

- Parents
- Peers
- Siblings
- Teachers
- Timed situations
- Specific mathematical concepts

Chapter 6 – "What Can I Do?"

- Targeting Maths skills and understanding
- Flipped learning approach
- Peer mentoring
- Cooperative learning
- Teacher impact
- Addressing avoidance
- Increasing self-awareness
- Breathing techniques
- Relaxation techniques
- Systematic desensitisation
- Expressive writing
- Psychodrama therapy
- Stories and storybooks
- Animal therapy
- Cognitive restructuring
- Reappraisal
- Mathematical mindset
- Mathematical resilience

Chapter 7 – "Can Teachers Be Maths Anxious, Too?"

- Teachers' Maths Anxiety
- Maths teaching anxiety
- The impact of teachers' Maths Anxiety

Chapter 8 – "Can We Solve the Equation?"

- Summary
- Looking to the future

Introduction

Maths Anxiety is such an interesting area of research, from both an educational and psychological perspective.

Educational researchers have long been interested in negative affect towards Maths and the role it has upon an individual's mathematical understanding and achievement, whether it be dwindling confidence as a student grows up, or a decline in enjoyment. Over the last 60-70 years in particular, there has been an exponential increase in the amount of Maths Anxiety research carried out, including how this might affect the teaching and learning of Maths in the classroom. At an educational level, whether research has focused on Maths Anxiety in the early years classroom, or at university level, findings have provided us with useful insights that can help us support students and embed positive practices in the classroom.

> Growing up, I was always pretty good at Maths. I didn't find it particularly hard, and at one point, I think when I was in Year 3, I loved it. My teacher used to give me 'tricky questions' to solve, and he would always make me feel like I had achieved something, every single lesson. If I got something wrong, he would help me figure out what happened, and I then had the desire to solve similar questions over and over, until I had mastered the concept. I loved it, and I loved being good at it. My next teacher, however, pointed out everything I got wrong, all the time. For some reason, I will always remember her telling me during one particular lesson, quite loudly in front of the class, that she found it ridiculous that I was getting the extremely hard questions right, but the basic ones wrong. Perhaps she thought I was being lazy, or copying someone else's work, but I honestly wasn't.
>
> Sadly, I think this was about the time where I stopped loving Maths. I did fine in my exams, and I was even placed in the top set in Maths in secondary school, but I quickly realised this wasn't the right setting for me. I was often made fun of for asking questions, or for not understanding a concept as quickly as others in the class, which always made me feel anxious and on edge. I avoided the class as often as I could and never spoke unless it was essential. Eventually, after a couple of years, I requested to move to the set below, as I was feeling so nervous within those four walls. I moved sets and remained there, doing fine at Maths and dropping it as a subject as soon as I could. I do wonder what would have happened if I hadn't had any negative experiences associated with learning Maths, but perhaps I wouldn't be so invested in helping those who are maths anxious today.
>
> My initial interest in Maths Anxiety was ignited during my first year of teaching, when a student in my classroom became avoidant, unmotivated and overwhelmed - all of which was strikingly different to their normally happy, engaged self. After continually trying, and failing, to understand what was happening, I started looking into the research that was available to me, which was surprisingly quite little at the time. Piecing existing research together with my own thoughts and observations, I came to the conclusion that they were experiencing Maths Anxiety. So, throughout the year, I worked hard to help them develop a more positive relationship with Maths and carried

on learning about this phenomenon as much as I could. While this student never became the most enthused Mathematician in the class, I do believe that their motivation improved, alongside their attainment and overall relationship with Maths.

Following this success, I was hooked. Maths Anxiety became the subject of my Master's research, as well as my Doctoral research, and my 'thing' in education was born. I began to share my research, findings and overall understanding of Maths Anxiety when delivering seminars, webinars, and teacher training, and loved how both teachers and researchers had Maths Anxiety as a shared interest, or perhaps more accurately, shared enemy.

–Dr Heidi Kirkland

Psychologists have also taken a keen interest in Maths Anxiety as it sits at the intersection between emotion and cognition. As you will see later in this book, psychologists often approach the study of Maths Anxiety from the point of view of wanting to examine psychological characteristics and behaviours that might contribute to the development, maintenance, or exacerbation of Maths Anxiety. Many of the strategies that psychologists propose for addressing Maths Anxiety focus on the way that students think and feel; this is then applied to the specific domain of Maths.

This is why we feel that our combined experience and understanding, as researchers and educators in various educational contexts, provides an excellent, well-rounded background to Maths Anxiety, which best equips you with all the information and practical advice you may need, right here in this book.

I was about eight years old, sitting in silence in my classroom. The entire class were stooped at their desks, scribbling in their Maths workbook, trying their best to complete the exercises set by the teacher. Every so often the teacher would reprimand one of my peers for talking, and proceeded to inform the class that they needed to complete the exercises before they could go outside and play. I knew I was capable of the calculations presented in front of me, yet for some reason my mind went blank. Everyone was immersed in their work, attempting to complete the task that had been set before it was time for a break. I was fairly good at Maths. Why then was I unable to concentrate on the task at hand?

I kept looking at the clock, then back at my work. Numbers would flash before my eyes, making sense of what they were was no effort, but deciding on what to do with them seemed an impossible task. The longer I went without writing anything, the more I began to panic, until eventually, the time was up. I sat on my own in the classroom with the teacher at her desk, while everyone else went to play. I knew that I hadn't been disruptive, so why did it feel like I was being punished? I'd spent the good part of an hour sitting in silence, trying my very best to complete the exercises, so why did the teacher think another thirty minutes would help? Asking myself these questions simply made me panic more.

After a few minutes, the teacher came over to me and declared that the problems were not even difficult, and if I were to just concentrate, I would get them done. Was it me? Even though I was trying to concentrate, was I really not? Was I inadequate in the skill of concentration? She pointed at a question and waited for me to answer it. I could smell the coffee on her breath and the impending build-up of frustration that I was not answering a question that she deemed to be so simple for someone of my ability. She asked why I couldn't do it, in a tone that implied frustration. I answered truthfully, 'I don't know', knowing full well that she would become angry by this. She did. I began to cry. She left me to work until break time was over and it was time for the rest of the class to come in and start some different work.

I do feel as though that teacher set me back in my Maths Education. And I now realise that she induced a maths anxious reaction in me. It wasn't until the following year when I had a more understanding teacher that I felt more comfortable and confident with Maths. I now wonder how many children have had a similar experience during their early school years. More importantly, how many of these experiences have hindered a child's schooling and development of mathematical skills, not only through childhood, but throughout adolescence and into adulthood as well? These are just some of the questions I have pondered since researching this important topic.

–Dr Thomas (Tom) Hunt

We hope that after reading this book you will feel more confident and supported to tackle any negativity towards Maths within your classroom, to better understand the complexity of Maths Anxiety, and to take action that is best suited to you and the students you teach, guide, and support. We also hope you enjoy reading it!

1 What Is It?

When people know that our research and passion focus on Maths Anxiety, and that we've worked on understanding it for a long time, they will probably ask one of two questions:

"What's that then?"

People often expect a perfectly primed answer, but the truth is it can vary greatly between individuals and the specific context, which makes defining it quite difficult.

"Is it real, like an actual thing?"

…yes! Maths Anxiety is related to the area of the brain associated with threat detection and pain itself. This shows how real Maths Anxiety can be, and how maths anxious individuals' brains can respond to Maths as though it is a physically painful experience.

> Lyons and Beilock (2012) investigated self-reported Maths Anxiety and brain activity. When anticipating a Maths task, they found that people with high levels of Maths Anxiety had increased activity in the bilateral dorso-posterior insula, which is associated with visceral threat detection and the experience of pain. Comparing this brain activity to when they were doing Maths, they found that the anticipation of Maths is more painful than actually doing the Maths itself.

Some people, including some teachers we have spoken to, find it difficult to acknowledge that anxiety about Maths exists. For many people, usually those who have experienced repeated success with Maths and feel confident in their abilities, there is nothing inherently scary about Maths. They have a point. Objectively, there isn't. However, the field of psychopathology tells us that many specific phobias exist in which there is a pronounced fear of situations or external stimuli that are often relatively harmless.

Existing classification manuals don't include specific reference to Maths Anxiety, though. One challenge with its inclusion could be that there is sometimes a vagueness associated with what Maths "is," as it is perhaps less clearly defined in comparison to typical objects or situations associated with phobias, such as spiders or heights. In fact, it is unlikely that someone will receive a diagnosis of a specific phobia of Maths. Instead, it is possible that another anxiety disorder offers a clinician a suitable diagnosis option for someone with high Maths Anxiety, such as Generalised Anxiety Disorder or Social Anxiety Disorder, but only if there is overlap between types of anxiety experienced. It remains to be seen whether commonly used guidelines for clinical diagnoses will come to include Maths Anxiety.

> The American Psychiatric Association's (2013) Diagnostic and Statistical Manual (DSM-5), a diagnostic system for mental disorders, lists "Specific Phobia" as a disorder, including several possible categories. A scan of the diagnostic criteria for specific phobias in the DSM-5 reveals similarity with the experience of Maths Anxiety: (i) unreasonable or excessive fear; (ii) an immediate anxiety response; (iii) avoidance (or extreme distress through endurance of the object/ situation); (iv) life-limiting; (v) a duration of at least six months; and (vi) not caused by another disorder.

In our opinion, it is possible that the symptoms of extremely high Maths Anxiety might be sufficient for classification of a specific anxiety disorder. However, Maths Anxiety is not listed as a distinct category of specific phobia in existing manuals, such as the American Psychiatric Association's Diagnostic and Statistical Manual (DSM-5) or the World Health Organization's International Classification of Diseases (ICD-11).

In the 16th century, John Napier, a Scottish mathematician, wrote the following nursery rhyme:

> "Multiplication is vexation, division is bad. The rule of three doth puzzle me, and practice drives me mad."

It wasn't until the 1950s though that anxiety in relation to Maths was researched and definitions began taking shape. In 1954, Mary Gough introduced the term "mathemaphobia," and, since then, Maths Anxiety has been increasingly researched, both in the education and psychology fields.

> Dreger and Aiken (1957) introduced the concept of "Number Anxiety," which developed over time to be more inclusive, into what we now refer to as "Maths Anxiety."

Maths Anxiety has also received media attention in the UK, from daily newspapers such as The Guardian or TV shows, such as Good Morning Britain. On top of this, there are also a range of organisations that focus on improving Maths Anxiety, or that have been developed specifically to help those who experience Maths Anxiety.

> ### Resources
>
> These organisations, groups, and people have explored Maths Anxiety, and some have ongoing initiatives to promote awareness and action. The resources offered may provide you with opportunities for further learning about Maths Anxiety:
>
> **The Maths Anxiety Trust** – www.mathsanxietytrust.com
>
> **Learnus** – www.learnus.co.uk
>
> **Steve Chinn** – www.stevechinn.co.uk
>
> **The Mathematics Anxiety Research Group** – www.marg.wp.derby.ac.uk
>
> **National Numeracy** – www.nationalnumeracy.org.uk
>
> **Mindful Maths** – www.mindfulmaths.org

While there is clearly increasing attention on both a public and research level, Maths Anxiety is still often misunderstood or remains hidden and unknown. The first article Heidi ever wrote, back in 2016, was titled *"Isn't it just a dislike for learning Maths?"* This was a quote from a previous colleague when she was discussing Maths Anxiety in a staff meeting, illustrating that not every teacher knows what Maths Anxiety is, or what's involved in the experience. If some teachers aren't aware of what Maths Anxiety is, then it is likely that many students and parents aren't either.

> Maths Anxiety Trust (2018) commissioned a survey of 2,019 adults in the UK. They found that 80% had never heard of the term Maths Anxiety.

Simply put, Maths Anxiety is a negative reaction to learning or using Maths. It is emotional, as it can involve feelings such as panic, fear, or apprehension. It is also related to cognition, such as attention, memory, and intrusive thoughts. It can also be seen physiologically, through bodily reactions such as sweaty palms, an increased heart rate, or changes in breathing. Behaviourally, Maths Anxiety is related to avoidance, and other behaviours associated with a negative attitude towards Maths or lack of motivation.

Maths Anxiety can be experienced in the classroom at any stage of education, or when people are no longer of a schooling age and school is a distant memory. It is not always obvious, sometimes it's even invisible. A black cloud of anxiety can hang over someone when they are working out the price of discounted items in a shop, or when they are trying to increase their recipe for 4 people to cater for their party of 12. It may happen regularly, or occasionally. It can be a mild feeling, or a full-on physiological reaction that consumes that person to their core and pulls the handbrake on what they're doing – it differs for everyone experiencing it.

Toby's Experience

Being compared to other people mathematically was Toby's self-proclaimed worst nightmare. He felt that he was OK at Maths, but he hated it and didn't see why it was useful to learn, especially because of easy access to calculators and AI these days. One trigger of his Maths Anxiety was when his teacher displayed the class scores on the whiteboard, for everyone to see. This comparison made him instantly feel apprehensive and overwhelmed and left him wanting to leave the classroom immediately. He felt that he was unmotivated because, when he was younger, he tried really hard to understand Maths, but now didn't see the point in this as he was never anywhere near the top of the class. Toby was certain he would not continue to learn Maths longer than he had to.

Something that adds a further level of intricacy in understanding Maths Anxiety is the difference between state anxiety and trait anxiety. State anxiety refers to the feeling of anxiety that a person has at the current moment in time. This makes no assumption about how anxious that person felt before, or how they might feel later on. As such, we might ask the question *"How anxious do you feel right now?"* On the other hand, trait anxiety refers to more consistent feelings of anxiety, and how a person may typically feel in general. On this basis, it would not be unreasonable to expect someone's anxiety next week to be similar to how they are feeling right now, assuming nothing significant has occurred in that time to induce greater or lesser anxiety. This still has the potential to change, but over a longer period of time. For example, a parent might notice a change in how anxious their child appears around Maths from one year to the next. This distinction is important, as it can not only influence how we interpret findings, but it can also impact what may be effective in reducing Maths Anxiety.

Beilock and Willingham (2014) note that some people assume Maths Anxiety means the same as being "bad" at Maths and that it's associated with poor Maths skills. However, they point out that Maths Anxiety is more than simply being bad at Maths, implying that a person could get better at Maths if they were not so maths anxious.

By now, you can probably already tell that Maths Anxiety is pretty complex, perhaps too complex to accurately define in one sentence. There are actually multiple definitions of Maths Anxiety which all differ slightly, but they do seem to have an air of similarity surrounding them.

A Few Definitions of Maths Anxiety

In their definition of Maths Anxiety, Richardson and Suinn (1972) referred to the tension and anxiety that people feel when manipulating numbers and solving mathematical problems in both ordinary life and academic situations. This definition of Maths Anxiety therefore opens up the possibility that Maths Anxiety can be experienced by

> many people in many situations; it is not necessarily in response to more challenging or complex Maths.
>
> Tobias and Weissbrod (1980) use the words panic, helplessness, paralysis, and mental disorganisation to describe Maths Anxiety. These are features of Maths Anxiety that some people instantly identify with and they seem to pertain to the emotional, behavioural, and cognitive aspects of Maths Anxiety. Interestingly, Tobias and Weissbrod's (1980) definition of Maths Anxiety also appears to relate more to the current state of a person, such as their reaction to Maths "in the moment."
>
> On the other hand, again in reference to the "ordinary" manipulation of numbers and solving of mathematical problems, Ashcraft and Faust (1994) describe how Maths Anxiety is associated with tension, apprehension, and dread. This definition implies an experience of anxiety that could occur ahead of actual mathematical problem-solving. Indeed, Maths Anxiety is often experienced in the build up to a situation involving Maths. Their use of the word "interfere" is also pertinent here. As we will discuss later in the book, Maths Anxiety has important cognitive features at its core, including the thoughts that people have but also the way it can impact the cognitive processes involved in mathematical problem-solving.
>
> Others have also offered definitions of Maths Anxiety, such as Trujillo and Hadfield (1999) who refer to the state of discomfort that might be experienced if an individual perceives a mathematical situation as a threat to their self-esteem. Once again, this appears to have a strong cognitive focus.
>
> It is also useful to note the American Psychological Association definition of Maths Anxiety, which broadly, and briefly, defines Maths Anxiety as the apprehensiveness and tension associated with one's arithmetic performance and performance on other mathematical tasks.

The majority of Maths Anxiety definitions involve the word "tension," referring to mental or emotional strain. Most of us can relate to this in one way or another, but it is important to acknowledge that, in this context, tension is only in relation to situations involving Maths. Further emotions or feelings are repeatedly mentioned, such as panic and apprehension, which can happen both in and out of the classroom. Whichever definition you prefer, it is clear that Maths Anxiety is a negative reaction, which has a negative effect on an individual.

Chapter 4 will look into the features of Maths Anxiety in a lot more detail, but on a basic level, the experience of Maths Anxiety may be associated with the following:

ATTAINMENT – the relationship between Maths Anxiety and attainment has been heavily researched. The usual assumption is that if someone struggles academically in Maths, they are more likely to experience Maths Anxiety. This may be more likely, but it is not always the case. There has also been a lot of debate over whether poor attainment is the cause of Maths Anxiety, or a result of it.

WORKING MEMORY – if someone is maths anxious, their anxiety can cloud their mind. This can make it challenging to meet the demands of the task at hand, as their working memory is over-exerted by their anxiety.

SPECIFIC MATHEMATICAL CONCEPTS - Maths Anxiety is not always a blanket reaction to Maths as a whole. Some areas of Maths are particularly anxiety-evoking for different people. These may include division, fractions or algebra, as a few examples. Alternatively, using and learning some concepts may instead elicit feelings of relief, acting as a comfort zone, such as procedural methods for calculations, or geometry. This is again likely to be unique to each individual though, related to their own beliefs, attitudes, attainment, and previous experiences.

SELF-BELIEF - typically, people with higher levels of Maths Anxiety will have lower self-belief when it comes to their Maths ability. We say "typically," however, because it is quite possible for someone to feel they are reasonably good at Maths, even better than average, but to still feel anxious about Maths. This is where previous experiences and specific context is important, such as the situation that a person is in, or relative judgements of one's ability in relation to those around them.

ATTITUDES - someone experiencing Maths Anxiety may be more likely to have negative attitudes towards using and learning Maths. For example, they may feel like learning Maths may be pointless, as it isn't useful in everyday life, or they may think it is boring or unenjoyable.

MOTIVATION - closely linked to negative attitudes, a maths anxious person may be more likely to have a lack of motivation, or be reluctant in situations involving Maths. This may be seen through a lack of enjoyment, feeling bored or apathetic, or even having learned helplessness, as a few examples.

AVOIDANCE - maths anxious individuals may be likely to avoid situations that involve Maths, which can occur at several levels. At the micro level, a student might avoid a specific Maths question on a test. At the macro level, they might avoid further study or jobs that involve, or are thought to involve, Maths. There can be various forms of avoidance in between, such as actively disengaging from what a teacher is saying, or putting off Maths homework.

OTHER PEOPLE - peers, parents, siblings, and teachers can all be influential when it comes to Maths Anxiety, through the language and behaviour they use. Social threats, including stereotype threat, can also be perceived by maths anxious individuals in situations involving others, such as friendly competitions involving Maths, or the comparison of mathematical abilities.

Yanuarto (2016) distinguished between two types of Maths Anxiety:

Global Maths Anxiety - when a person has an anxious reaction in all situations involving Maths, regardless of what the situation or content is.

Specific Maths Anxiety - when a person has an anxious reaction to certain situations involving Maths, but not every situation.

This is similar to the trait and state definitions of anxiety that we discussed earlier in this chapter and supports the idea that not every maths anxious individual will have the same blanket reaction to Maths at the same intensity; it does vary.

All of these features may be present in an individual's experience of Maths Anxiety, or hardly at all. They may be identified through observable behaviours, although some may prove trickier to spot. The features we've just highlighted are by no means a comprehensive list and should not serve as a guide to understanding what everyone's experience is like; they just provide a good starting point for us to understand what may be part of the experience, so that we can help those who are maths anxious. Ultimately, the more research and conversations there are about the experience of Maths Anxiety, the more we can learn and act accordingly.

> **Joanne's Experience**
>
> Finding learning in general quite challenging, Joanne often shared that Maths was "the worst." It was so hard, and she felt that she couldn't do anything right. Never answering questions or getting involved in class discussions was one way that Joanne tried to escape feeling anxious, but she always felt the dread of being asked a question by the teacher which she had no clue how to answer. This thought even crept into her dreams, especially about questions relating to fractions.

Perspectives of Maths Anxiety

How Maths Anxiety is researched and perceived is important to discuss at this point. A lot of existing research mainly looks at specific factors of Maths Anxiety, rather than the whole experience itself. For example, a researcher may closely examine the relationship between self-belief and Maths Anxiety through testing a hypothesis, such as:

> "The higher the level of Maths Anxiety, the lower the level of self-belief an individual will have."

This type of objective research is really useful to other researchers and teachers, because it gives us a detailed look into specific relationships and aims to give measurable, numerical information that we can potentially act upon. To achieve this though, investigations may be conducted outside of the classroom or "real-life situation," such as in a research laboratory, where these variables may be manipulated, measured, and examined. This brings into question whether research data is replicable and reflective of real life.

> Arslan et al. (2015) looked at the research methods used in 40 studies that investigated Maths Anxiety, published between the years 2000 and 2013. They found that 33 out of 40 were quantitative and based on numerical findings. Only four were qualitative, and used methods such as interviews and observations to gain data, while only three used mixed methods.

Similarly, the collection of data may be retrospective, where adults are asked how they felt when they were a child, for example. This muddies the water, as you can imagine. It is hard for most people to remember something a few years ago, let alone remember back to their primary school years. Therefore, it is important to consider the effects of vague memories, or skewed memories, upon any subjective recounts collected within a retrospective study. We also need to question what is really being measured if Maths Anxiety is only assessed in a single moment in time, often called "snapshot studies." This also relates back to the distinction between state and trait anxiety.

Sometimes, objective research can reduce Maths Anxiety down to something that somebody "has," with causes and symptoms. It can be a label that is given to people. This is a simplistic view of Maths Anxiety, because as you will see in later chapters, identifying features of a Maths Anxiety experience does not necessarily indicate definitive causes of it – it is much more complex and multifaceted than that.

It is our view that the experience of Maths Anxiety is individual to each person, including how it manifests and what works when trying to address it. That said, there appear to be several commonalities between experiences of different people. Also, as we will discuss, certain behaviours and psychological variables, such as attitudes or beliefs for example, seem to predict, and be predicted by, Maths Anxiety quite well. Given the wide range of factors connected with Maths Anxiety, it is important to approach things holistically.

Think of it like a pair of tangled headphones.

If you work on only one knot, your headphones are very unlikely to unravel. If you focus on one aspect of Maths Anxiety, you are unlikely to fully understand what the individual is experiencing.

If you focus on the entirety of the headphone wires and work out how they are interconnected and tangled, you will be more likely to unravel them with success. If you look at the whole experience of Maths Anxiety, you are much more likely to understand what the individual is going through, and what their experience is like.

Keeping with the same analogy, in a similar way to how the headphone wires could have multiple knots, a person's current level of Maths Anxiety might be based on a series of challenges and negative experiences they have faced. Alternatively, they might have had a particularly traumatic or troublesome experience that requires even more careful and diligent unravelling, much in the same way that one large knot requires time, patience, and initial planning.

> Stoehr (2017) investigated whether there were common features across three female trainee teachers' experiences of Maths Anxiety. She found that, while there were commonalities between experiences, each woman had specific and distinct fears in relation to Maths. This shows how it is important to understand each individual's experience of Maths Anxiety, because they are unique in their own ways.

Maths Anxiety Vs Other Anxieties

Having explored the basics of Maths Anxiety, you might now wonder if maths anxious people are anxious in other areas of their lives too. Possibly, but not necessarily. When Heidi was researching Maths Anxiety for her PhD, one of her participants was very maths anxious but was confident in all other aspects of her life. She played in piano concerts, put herself forward for important school roles, and performed in national ballet performances. When she was asked to answer a question in her Maths lesson, she froze and began to panic. Her reaction, like that of many others, was specific to Maths.

Some researchers have compared Maths Anxiety to other forms of anxiety, such as General Anxiety (persistent anxiety and worry that is experienced across a range of situations), Test Anxiety (experienced in response to exams or class tests), and anxiety in other subjects. Through comparing these, research has been able to show that Maths Anxiety is unique, even though it overlaps with other types of anxiety. This means that although there are similar features that can occur in both Maths Anxiety and other anxieties, it is distinct enough to be classified in its own right.

However, the consistent statistical relationship that researchers observe between Maths Anxiety and General Anxiety should not be ignored. There is clearly some overlap and one argument that is put forward is that being generally anxious predisposes a person to experiencing Maths Anxiety. More longitudinal studies, especially on a large scale, are needed to test this, although there is some evidence to suggest that students' anxiety profiles change between primary and secondary school.

> Hill et al. (2016) wanted to explore the relationship between Maths Anxiety and General Anxiety in primary and middle school children. They found that the two types of anxiety were moderately related in a statistical sense but were ultimately independent from each other. However, they did find that partialling out (statistically controlling) the effect of General Anxiety reduced the significant negative relationship initially observed between Maths Anxiety and Maths achievement.

People experiencing extremely high General Anxiety, perhaps even having been diagnosed with General Anxiety Disorder, are likely to have similar physiological and emotional reactions to a person experiencing Maths Anxiety. For example, both may have increased heart rates, perspiration, and feelings of apprehension and tension. However, Maths Anxiety is specific to contexts involving Maths – it is domain specific. On the other hand, General Anxiety refers to anxiety experienced across a multitude of situations, and it is often difficult to ascertain its onset.

> Carey et al. (2017) measured General Anxiety, Test Anxiety, and Maths Anxiety in 1,720 students in Year 4, Year 7, and Year 8 in England. They found that the younger children's anxiety types did not differentiate. That is, children's anxiety was generally low, slight, moderate, or high across all anxiety types. However, by secondary school, different anxiety profiles exist, including some students who only score highly on measures of academic-specific anxiety, i.e. Maths Anxiety and Test Anxiety.

It has also been argued that the lines are blurred between Maths Anxiety and Test Anxiety. As part of their experience, a maths anxious individual can respond negatively to timed situations and completing tests. When asked about one of their most nerve-wracking situations in school, adults who are maths anxious will often recount their memories of tests where they felt very pressured. This shows that maths anxious people are likely to experience anxiety in association with tests, but these are often specific to Maths tests and timed situations involving Maths, whereas Test Anxiety is not subject specific and is relevant to test situations more generally. Therefore, even though there is an overlap, Maths Anxiety is still a distinct experience that cannot be accounted for by Test Anxiety alone. There is research evidence to support this, with several studies reporting a significant correlation between Maths Anxiety and Test Anxiety, demonstrating consistency in the relation between them. Importantly, the correlations are not strong enough to suggest that they are essentially the same thing, with a correlation of $r = 0.5$ probably being quite typical (a moderate positive correlation).

> ### Ira's Experience
>
> When completing Maths tests, Ira often became extremely anxious. These ranged from end of unit tests, practice papers for official exams, or weekly low-stakes reviews on what she had learnt that week. She didn't see these as low-stakes though, as she shared that her heart was always racing and her mind just went blank. She felt that everything that she had learnt, everything that she thought she knew, was no longer in her mind.

There are of course other subjects that elicit anxiety as well. Subject-specific anxieties can create the same negative responses as Maths Anxiety, but again, they are related to specific situations only. For example, if someone is experiencing Literacy Anxiety, those individuals experience anxiety in situations where they are learning about, or using the skills of, Literacy, not Maths. Other subject-specific anxieties are not quite as researched to the degree that Maths Anxiety is though. Perhaps this is because of the (often perceived) right or wrong nature of Maths, or the fact that it has a bad stigma and is perceived as a tricky subject that people either love or hate.

> Punaro and Reeve (2012) aimed to compare Maths Anxiety and Literacy Anxiety in 58 nine year olds. The participants were given a judgement task focusing on 48 Maths and Literacy problems and were asked to rate their worry using a picture scale. This was a Likert scale that used pictures, like emojis, to show faces with different degrees of worry, ranging from extremely worried, to not worried at all. They reported that both subjects elicited some level of anxiety, but anxiety towards Maths was more severe.

Researchers have also found anxiety specific to other subjects, such as anxiety in the performing arts, Reading Anxiety, or Science Anxiety. If someone is anxious across all these subjects, it may be more likely that they are experiencing generalised anxiety towards learning or the school environment, rather than specific anxiety towards a certain area, like Maths Anxiety.

So… What Is It?

- Definitions of Maths Anxiety vary.
- Maths Anxiety is a negative reaction to learning or using Maths.
- Conceptualising and defining what Maths Anxiety actually is has improved greatly, but more needs to be done, as it can sometimes be misunderstood and inappropriately generalised.
- Attainment, working memory, specific Maths concepts, self-belief, attitudes, motivation, avoidance, and other people can all be factors associated with the experience of Maths Anxiety.
- Maths Anxiety is a fluctuating experience that is unique to each individual.
- Maths Anxiety is different to other forms of anxiety, such as General Anxiety and Test Anxiety, but there is some overlap and students' anxiety profiles may change as they progress through the education phases.

Hopefully, you now feel that you have a more rounded view of what Maths Anxiety is, and importantly, what it isn't. If someone asks you what it is, or if it's real, you should now feel equipped to tell them to pull up a chair before you dive deep into the realm of Maths Anxiety and all of its complexities.

There are a few important things to reflect on from this chapter, before reading on. Even though there has been increasing research and various definitions since the 1950s, leading to a deeper understanding of the experience, Maths Anxiety is still sadly a part of the classroom, and extends beyond into everyday life. We truly believe that each person's experiences in life are unique and personal to them, because of how they have been shaped by their experiences, cognition, emotions, and beliefs. Maths Anxiety is no different – it is unique to the person experiencing it. Even though there are commonalities across experiences and things to look out for, or observable behaviours that may highlight that someone is maths anxious, the makeup of their experience is theirs. Therefore, the first thing you need to do to help someone experiencing it, even if it's yourself, is to understand that experience on an individual level.

References

American Psychiatric Association, DSM-5 Task Force. (2013) *Diagnostic and Statistical Manual of Mental Disorders: DSM-5*. 5th edition. Vancouver: American Psychiatric Publishing Inc. Available at: https://doi.org/10.1176/appi.books.9780890425596.

American Psychological Association. (n.d.) 'Mathematics Anxiety', In *APA Dictionary of Psychology*. Available at: https://dictionary.apa.org/mathematics-anxiety (Accessed: 28 April 2024).

Arslan, C., Deringol-Karatas, Y., Yavuz, G. and Erbay, H. N. (2015) 'Analysis of Research on Mathematics Anxiety in Selected Journals (2000-2013)', Procedia - Social and Behavioral Sciences, 177, pp. 118-121. Available at: https://doi.org/10.1016/j.sbspro.2015.02.355.

Ashcraft, M. H. and Faust, M. W. (1994) 'Mathematics Anxiety and Mental Arithmetic Performance: An Exploratory Investigation', Journal of Cognition and Emotion, 8(2), pp. 97-125. Available at: https://doi.org/10.1080/02699939408408931.

Beilock, S. L. and Willingham, D. T. (2014) 'Math Anxiety: Can Teachers Help Students Reduce It? Ask the Cognitive Scientist', American Educator, 38(2), p. 28.

Carey, E., Devine, A., Hill, F. and Szűcs, D. (2017) 'Differentiating Anxiety Forms and Their Role in Academic Performance from Primary to Secondary School', PloS one, 12(3), pp. 1-20. Available at: https://doi.org/10.1371/journal.pone.0174418.

Dreger, R. M. and Aiken, L. R. (1957) 'The Identification of Number Anxiety in a College Population', Journal of Educational Psychology, 48(6), p. 344. Available at: https://doi.org/10.1037/h0045894.

Gough, M. F. (1954) 'Mathemaphobia: Causes and Treatments', The Clearing House, 28(5), pp. 290-294.

Hill, F., Mammarella, I. C., Devine, A., Caviola, S., Passolunghi, M. C. and Szűcs, D. (2016) 'Maths Anxiety in Primary and Secondary School Students: Gender Differences, Developmental Changes and Anxiety Specificity', Journal of Learning and Individual Differences, 48, pp. 45-53. Available at: https://doi.org/10.1016/j.lindif.2016.02.006.

Lyons, I. M. and Beilock, S. L. (2012) 'When Math Hurts: Math Anxiety Predicts Pain Network Activation in Anticipation of Doing Math', PLoS ONE, 7(10), pp. 1-6. Available at: https://doi.org/10.1371/journal.pone.0048076.

Maths Anxiety Trust (2018) Official Figures. Available at: https://mathsanxietytrust.com/official-figures.html (Accessed: 17 May 2024).

Punaro, L. and Reeve, R. (2012) 'Relationships Between 9-Year-Olds' Math and Literacy Worries and Academic Abilities', Child Development Research, 2, pp. 1-12. Available at: https://doi.org/10.1155/2012/359089.

Richardson, F. C. and Suinn, R.M. (1972) 'The Mathematics Anxiety Rating Scale: Psychometric Data', Journal of Counseling Psychology, 19(6), p. 551. Available at: https://doi.org/10.1037/h0033456.

Stoehr, K. J. (2017) 'Mathematics Anxiety: One Size Does Not Fit All', Journal of Teacher Education, 68(1), pp. 69-84. Available at: https://doi.org/10.1177/0022487116676316.

Tobias, S. and Weissbrod, C. (1980) 'Anxiety and Mathematics: An Update', Harvard Educational Review, 50, pp. 63-70. Available at: https://psycnet.apa.org/doi/10.17763/haer.50.1.xw483257j6035084.

Trujillo, K. and Hadfield, O. (1999) 'Tracing the Roots of Mathematics Anxiety Through In Depth Interviews with Preservice Elementary Teachers', College Student Journal, 33(2), pp. 219-233.

World Health Organization (2019) International Statistical Classification of Diseases and Related Health Problems. 11th Edition. Available at: https://icd.who.int/en; (Accessed: 19 May 2024).

Yanuarto, W. N. (2016) 'Teachers' Awareness of Students' Anxiety in Math Classroom: Teachers' Treatment VS Students' Anxiety', Journal of Education and Learning (EduLearn), 10(3), pp. 235-243. Available at: https://doi.org/10.11591/edulearn.v10i3.3808.

2 Who Has It?

In 1954, Mary Gough suggested that Maths Anxiety is almost as common as the common cold. Today, if the topic of Maths Anxiety comes up in a group discussion, we can pretty much guarantee that someone will say "*I have that!*", or they will know someone who they think has it. We should be quick to say here though, that Maths Anxiety is not binary – it's not the case that you have it or you don't. However, many people will claim that they are maths anxious, which implies they have put themselves into the category of being a maths anxious person. Similarly, many teachers might tell us that they have a maths anxious pupil in their class.

Throughout this book, referring to someone as maths anxious is intended to mean someone who reports or displays a certain degree of Maths Anxiety. What that exact level is is not clear, but the language helps with describing what many people, including teachers, students, and the general public, understand to be the case.

How many people experience Maths Anxiety though? Are there any potential trends in who experiences it, or can it be experienced by anyone? Knowing how widespread a problem Maths Anxiety is can be quite challenging. The prevalence, meaning how common it is, varies hugely depending on who you talk to, or what you read. Putting Maths Anxiety into context can therefore be tricky.

The Prevalence of Maths Anxiety

In our opinion, many people experience Maths Anxiety.

> Yanuarto (2016) suggests that the number of people who are affected by Maths Anxiety is "overwhelming."

Many statements surrounding the facts of how many people experience Maths Anxiety are somewhat vague, so it is useful to have quantitative data to properly understand the proportion of the population that experience Maths Anxiety. There are several data sources out there that report on the prevalence of Maths Anxiety. Some researchers may claim that a certain percentage of people experience Maths Anxiety in relation to a very specific population, such as first-year undergraduate students in the UK, while some may be more vague, referring to adults in general. The reported prevalence of Maths Anxiety varies quite a lot, from 2% all the way to 85%, so it is important to look at the context of the research. Who participated in the study, and how many participants were there altogether? Where was it conducted, and how was the information collected?

Looking specifically at the occurrence of Maths Anxiety in the UK, the Maths Anxiety Trust commissioned two nationwide surveys in 2018 to explore what proportion of the UK population experienced Maths Anxiety. The findings were explored at the Learnus UK 2018 Maths Anxiety Summit, where researchers and educators gathered to discuss the extent of the problem in the UK. Here, it was reported that in one survey, 20% of adults in the UK had felt maths anxious at some point in their lives, while 8% of those adults had felt this way within the last month. Comparatively, 36% of 15-24 year olds said in the survey that they had felt maths anxious at some point previously. This was taken from the general population with 2,019 participants. The second survey had a similar number of participants, but this time they were aged between 11 and 18 years old and attended high-performing schools. They reported that 40% of students were maths anxious sometimes, while 25% were anxious most of the time.

More recently, in 2023, the charity National Numeracy, commissioned by KGMP UK, asked 3,000 adults in the UK questions related to their attitudes and feelings towards Maths. They found that 35% of participating adults said that doing Maths makes them feel anxious, while 29% said that they actively try to avoid mathematical situations. In fact, 52% of them recalled that they stopped studying Maths as soon as they were able to, while 20% reported that doing Maths made them feel physically sick due to the anxiety and panic they felt.

These surveys highlight that yes, Maths Anxiety is an issue in the UK, and yes, it affects a notable proportion of the population.

A Glance at Statistics

Burns (1998) suggested that 2/3 of Americans fear and loathe Maths.

Thilmany (2009) found that 60% of university students had experienced Maths Anxiety.

Richardson and Suinn (1972) identified that 11% of their university sample had high levels of Maths Anxiety that warranted counselling.

Ashcraft and Moore (2009) estimated that approximately 17% of the entire population is high in Maths Anxiety, based on the trend of their collected data.

Betz (1978) found that 68% of students in college Maths classes had high levels of Maths Anxiety.

Perry (2004) reported that 85% of students who were enrolled in introductory Maths classes for their college course experienced at least mild anxiety.

Ashcraft and Ridley (2005) found that 20% of university students experienced high Maths Anxiety.

Chinn (2009) identified between 2% and 6% of 2,084 secondary school students in the UK experienced high levels of Maths Anxiety.

> Johnston-Wilder, Brindley and Dent (2014) shared that 30% of 226 apprentices in the UK experienced high levels of Maths Anxiety, while 18% experienced it less intensely.
>
> Jones (2001) had a participant pool of 9,000 students and found that 25.9% had a moderate to high need for help with their experience of Maths Anxiety.
>
> Hart and Ganley (2019) studied 1,000 people in the general population and found their self-reported Maths Anxiety to be normally distributed, with the average between "some" to "moderate."
>
> Karim, Hajar, and Tunku Ahmad (2023) stated that less than 1/3 of 138 secondary students had experienced Maths Anxiety.

You can see that the reported prevalence of Maths Anxiety differs widely. How can we say who is experiencing Maths Anxiety, when all the evidence is conflicting and confusing? Knowing the size of the problem, and how it's reported, can depend on many things, such as how Maths Anxiety is measured, or how it's defined. On top of this, there are many things that may be influential when reporting who is experiencing Maths Anxiety, such as gender, culture, educational background, age, and socio-economic status.

Firstly, reported prevalence will differ from study to study depending on how it is actually measured. We will discuss this in more detail in Chapter 3, but it is usually measured through questionnaires, sometimes through observation of behaviours, and occasionally indirectly via physiological or brain activity. The type of data derived from these different approaches will vary greatly, and reporting on the prevalence and extent of Maths Anxiety is often dependent on subjective interpretation, but also practical considerations such as the number of people being studied.

Let's look at an example. Say that we were measuring teacher satisfaction. Your school decided to use a questionnaire to measure how satisfied the teaching staff were, while our school wanted to measure satisfaction by observing how many times teachers smiled each day. The two different ways of measuring are most likely going to have different results, as they have used different methods to gather their data. Your school has very low satisfaction, because the data from your questionnaire illustrated quite a negative response. This means that the prevalence of satisfaction in your school is low. Our school has a higher level of satisfaction because we counted a lot of smiles, so our prevalence of satisfaction is high. Comparing the two like for like won't create the most accurate comparison. Similarly, comparing reports of prevalence for Maths Anxiety from findings based on different research methods won't lead to an accurate comparison either.

Discrepancies between reports of prevalence can also happen if both studies use questionnaires to measure Maths Anxiety though. There are plenty of questionnaires available to measure Maths Anxiety, for adults and for children. There are lots of similarities between them, but they are all inherently different due to the questions included. If one group of people complete a questionnaire to measure their Maths Anxiety in mathematical situations involving learning Maths, while another group completes a different questionnaire that focuses on being tested in Maths, it is probable that the results will show different levels of Maths Anxiety.

Let's look back at our example. Your school's questionnaire to measure teacher satisfaction focuses on work hours and the ability to progress in their career. Our school also used a questionnaire to measure the satisfaction of teachers this time but focused on opportunities for professional development and the quality of leadership. Even though we have now all used questionnaires, they are not focusing on the same areas relating to teacher satisfaction, so our findings are unlikely to be the same. This is also true if we were all to use questionnaires to measure Maths Anxiety – the chosen questionnaires may be similar but the questions are different, so our results and conclusions regarding prevalence could be too.

Another difficulty when trying to determine the prevalence of Maths Anxiety is understanding the different perceptions of researchers. As discussed in the last chapter, there are lots of definitions of Maths Anxiety. These again have underlying similarities, but each differs in their focus. If one researcher identifies individuals who are maths anxious by using one definition, focusing on physiological responses for example, while another researcher uses an alternative definition, which emphasises the role of emotion, this might affect the data collected which, in turn, could impact conclusions about how many people are considered to be maths anxious.

For example, if you believed that teacher satisfaction meant that teachers are content in their job, and this was measured, while we believed teacher satisfaction meant that teachers felt valued and therefore measured satisfaction in this way, we are essentially measuring different, although perhaps overlapping, concepts. This will undoubtedly affect the reported prevalence in our schools, as it would if we were utilising differing definitions of Maths Anxiety and reporting prevalence based on this.

Similar to varying definitions affecting prevalence, researchers are likely to differ in their perception of what "high" Maths Anxiety is. If we told you that in our school, 64% of teachers had high teacher satisfaction, your interpretation of what "high" is would probably differ to mine, and you would probably question the vagueness of this statement.

When reading various reports about the prevalence of Maths Anxiety, there are claims of "high levels of Maths Anxiety" or "moderate to high Maths Anxiety." It is often unclear what is meant by "high." Could this be someone who, for example, scores over 20 points out of a possible 30 on a Maths Anxiety questionnaire? Or could it be that a person scores a certain number of standard deviations higher than the sample mean? To complicate matters further, we should bear in mind that a sample is deemed to be a selection of people taken from a wider population, so the average Maths Anxiety score in a given sample is unlikely to match the "true" average within the population. These considerations greatly impact the way in which results from Maths Anxiety questionnaires are interpreted, ultimately affecting the conclusions people draw about the prevalence of Maths Anxiety.

The final issue we want to highlight when determining prevalence, and what may affect it, is the samples (groups of people) that are tested. Some samples may only include adults, while some samples might include students in Key Stage 2. Alternatively, one sample may be from Australia, while another sample is taken from Turkey. On top of this, there could be a question mark over the suitability of a particular Maths Anxiety questionnaire for use in one group of people compared to another, for example, due to the questionnaire's language and cultural context. The size of the sample also may impact the reported prevalence, with smaller samples potentially being less representative of the populations from which they are derived. Within these samples, there are further things that might need accounting for, such as how many females and males are in the sample, or how many people are voluntarily on a course that involves Maths, compared to those who aren't. Such samples aren't comparable, because they have underlying differences, which includes different levels of self-reported Maths Anxiety.

If we were to measure teacher satisfaction using the same questionnaire as each other, but your sample only consisted of Newly Qualified Teachers, while our sample only consisted of teachers who have taught for ten years or more, you wouldn't be surprised that our findings, and prevalence of satisfaction, would be different. This is similar to the prevalence of Maths Anxiety; it is very dependent on the group of people being measured.

This leads to the question of whether certain people are more likely to experience Maths Anxiety than others, which can impact reported prevalence.

Who Is More Likely to Be Maths Anxious?

One of the things that struck us the most when we began researching Maths Anxiety was that certain groups of the population are more likely to report higher levels of it. From the country you live in or your socio-economic status, to your genetic makeup or the age or gender you are, lots of researchers have suggested certain people are more likely to experience Maths Anxiety in their lifetime compared to others.

Geography, socio-economic status, genetics, age, and gender are primarily the areas that some researchers believe impact the likelihood of someone experiencing Maths Anxiety. Some researchers do not believe this to be the case though, suggesting that no one is predisposed to developing Maths Anxiety. Is there any truth behind these claims? Let's explore the most widely held beliefs as to who is more likely to be maths anxious than others.

Maths Anxiety Is Different in Each Country

Even though Maths is a universal subject, each country has its own educational policies, guidelines, curriculum, and general way of doing things. The emphasis and importance placed on learning and understanding Maths within those countries also differ as part of the embedded culture. Learning Maths may be universal, but is Maths Anxiety?

> Eccius-Wellmann et al. (2017) compared Maths Anxiety among Mexican and German students and found that German students had higher levels of Maths Anxiety.

The majority of research into the occurrence of Maths Anxiety comes from WEIRD countries; Western, educated, industrialised, rich, and developed. This has largely included the US and Western Europe, while less is known about Maths Anxiety in the global South. However, there are pockets of research carried out in specific countries and it is important for researchers and educators to seek out these reports and research papers to get a fuller understanding of the relevance of country-specific contexts and to identify differences and commonalities in approaches and findings.

> An OECD (2015) report showed that out of all the countries that took part in the Programme for International Student Assessment (PISA) 2012 survey, over half of 15 year olds reported that they often worry that it will be difficult for them in Maths classes. Over a third reported that they get very tense when they have to do Maths homework and that they get very nervous doing Maths problems.

The disproportionate amount of Maths Anxiety research carried out in WEIRD countries compared to others makes it difficult to make accurate comparisons of Maths Anxiety across countries, but organisations such as the Organisation for Economic Co-operation Development (OECD) work towards comparing attitudes and emotions towards Maths, globally. The OECD collects information across the world about students' attainment in reading, Maths, and Science, including other measures such as self-beliefs and attitudes pertaining to these core areas. It does this through its PISA. Since 1997, the survey has collected and provided comparative data from an increasingly large number of countries (over 80 most recently), on a three-year cycle. The caveat to the OECD data on Maths Anxiety is that the focus is on a specific age group, with students aged 15 years old. That said, there is a lot of useful data that helps us understand the global picture of Maths Anxiety.

> OECD (2023) published the data from their 2022 PISA, in which approximately 690,000 15-year-old students from 81 countries took part. The key findings related to Maths Anxiety are:
>
> - Students with higher Maths performance had, on average, lower levels of Maths Anxiety compared to students with poorer Maths performance. This was found in every participating country without exception.
> - Maths Anxiety was found to be higher in countries with lower Maths performance.
> - Finland, Netherlands, and Hungary had the lowest Maths Anxiety scores.
> - Guatemala, Brunei Darussalam, and Paraguay had the highest Maths Anxiety scores.
> - Hong Kong, Japan, Macao, and Chinese Taipei had high levels of Maths Anxiety but had high Maths performance ratings overall.
> - On average across participating countries, students with a growth mindset experienced less Maths Anxiety compared to students with a fixed mindset.

Taking all of these findings into account gives us a clearer picture and shows that there are both similarities and differences across countries in terms of Maths Anxiety.

It has been suggested that high-performing countries may report a low occurrence of Maths Anxiety overall because most of the students are academically strong, yet an alternative theory to this is that these countries could be rated highly in relation to Maths Anxiety because of the associated pressure and cultural demands to perform. For example, research has shown that Maths Anxiety may be prevalent in Asian countries due to the pressure of formal testing, fear of failure, and high expectations placed on students' achievement, but this could instead be related to the curriculum or education systems.

> Fan, Hambleton, and Zhang (2019) looked specifically at PISA data related to Maths Anxiety, comparing it in three countries, Finland, South Korea, and the US, due to their varying cultures and different mathematical performance. They reported that all countries reported Maths Anxiety, but the makeup of this in each country was different. They found that the country with the most cases of high Maths Anxiety was the US, while Finland had the lowest prevalence of high Maths Anxiety.

Reflecting on all of the work researchers have done to investigate Maths Anxiety across the world, we believe that Maths Anxiety is a universal experience, but also something that varies from country to country. The reasons behind this may be related to embedded culture, Maths attainment, available resources, or educational policies, but without further exploration, it can't be said with certainty.

Socio-economic Status Influences Who Is Maths Anxious

Socio-economic status refers to the social standing or class of someone, based on a combination of their education, income, and occupation. It is most commonly defined by combining parental educational level, career, and income and is usually broken down into high, middle, or low. People or families with a higher socio-economic status tend to have more money and disposable income and resources, resulting in higher privilege, power, and control. But how does this relate to Maths Anxiety?

> Short and McLean (2023) reported on Maths Anxiety and socio-economic status in 421 children in early years education in Scotland. They found that Maths Anxiety of high socio-economic status children fell during the first three years of formal schooling, whereas it increased for those of low socio-economic status.

This area is comparatively less researched than others, but it has been suggested that socio-economic status may impact the likelihood of people experiencing Maths Anxiety. Specifically, if someone has a low socio-economic status, it is suggested that they are more likely to experience Maths Anxiety compared to someone from a more advantaged background.

> OECD's (2013) PISA data from 2012 showed that differences in Maths Anxiety were less pronounced in relation to socio-economic status, compared to other areas such as gender, but there were differences nonetheless. In Greece, Bulgaria, Denmark, Singapore, and Liechtenstein, there was a particular disparity according to socio-economic status, as those who were more disadvantaged experienced higher levels of Maths Anxiety compared to those who were more advantaged. In Greece, 81% of disadvantaged students reported feeling mathematically anxious, compared to 63% of advantaged students. Similarly, in Singapore, 70% of students from a low socio-economic background felt maths anxious, while only 46% of students from a high socio-economic background reported that they were experiencing, or had experienced, Maths Anxiety.

The link between socio-economic status and Maths Anxiety may also be explained or moderated by other factors, such as attitudes and access to education. For example, if a student is not able to attend a high-performing school, or any school at all, their mathematical understanding may be lower, potentially affecting how anxious they feel about Maths. Many studies have linked poor mathematical performance with low socio-economic status, so this may be a logical assumption.

> Guzman, Rodriguez, and Ferreira (2021) explored the relationship between levels of Maths Anxiety and the socio-economic status of 451 second-grade students in Chile, with an average age of 5 and a half years old. They reported a direct effect of socio-economic status upon levels of Maths Anxiety, meaning that the higher the student's economic status was, the lower their levels of Maths Anxiety were, and vice versa.

Some studies have specifically focused on one area of socio-economic status; parental professions. Research has shown quite consistently that students whose parents with occupations that fall in the "professional" categories, related to areas such as health, teaching, science, or business, had higher motivation to learn Maths, better attitudes towards Maths, and overall, better academic performance in this area. Even though the explanation behind this is unclear, it stands to reason that this may also be linked to experiencing Maths Anxiety. Studies have investigated the role of parental occupation and its potential relationship with Maths Anxiety and found that students with parents who have a "professional" occupation, as a group, have lower reported prevalence of Maths Anxiety compared to those who don't.

> Lane (2017) compared the levels of Maths Anxiety in 356 students, aged 15-18 years old, in relation to their parents' occupations in Ireland. They found that the students with the lowest levels of Maths Anxiety had parents who worked with Maths. In comparison, students with the highest levels of Maths Anxiety had parents whose jobs were classified as "skilled" or "other" for their profession.

Socio-economic status, including parental occupation, is somehow related to the occurrence of Maths Anxiety. Why this happens remains unclear, but it appears that those with a lower socio-economic status are more likely to experience Maths Anxiety.

Melanie's Experience

Although there are definite trends in the prevalence of Maths Anxiety, and who may be more likely to experience Maths Anxiety than others, it's not a blanket truth. Take

> Melanie, for example. Melanie experienced Maths Anxiety throughout her education, even though her father was what would be classified as a "professional," and her mother was a secondary Maths teacher. This shows that these assumptions aren't steadfast rules that must be abided by; anyone can be maths anxious.

Genetics Can Determine Whether Someone Will Be Maths Anxious

For a long time, there has been debate about the role of genetics and the environment in determining certain anxiety disorders. This "nature vs nurture" debate has also existed in relation to Maths Anxiety. Does someone's DNA make it more likely that they will experience Maths Anxiety than someone else?

There have been a few investigations into the role of genetics in Maths Anxiety, to see if anything predisposes people to be more likely to be maths anxious than others. This is mostly done by looking at twins, and how they respond to learning and using Maths.

> Wang et al. (2014) measured Maths Anxiety in 514 pairs of twins and followed their journeys through education over seven years. They reported, through univariate and multivariate behavioural genetic modelling, that 40% of the variance in Maths Anxiety could be explained by genetics. The other 60% could be explained by environmental factors.

It is quite unlikely that genetics account for the entire Maths Anxiety experience, but they may have a role in predisposing someone to be maths anxious. It may be more likely that genetics has a role in anxiety more generally, as well as mathematical cognition, which includes how someone learns Maths, and therefore may indirectly affect who may be likely to experience Maths Anxiety.

> ### Aanya's Experience
>
> While Aanya doesn't have a twin, she shared that she often felt like "the odd one out" in her family. She saw her parents and siblings as being amazing at Maths and felt that they really enjoyed it. Aanya, on the other hand, felt maths anxious and didn't enjoy Maths in the slightest. Does this oppose research into the role of genetics, or could it highlight the influence of Aanya's environment?

Some researchers have indicated that it is possible for Maths Anxiety to be heritable from our genes, and that a third of the differences between maths anxious individuals and those who are not maths anxious can be explained by DNA. The rest of the differences that are unaccounted for by genetics are likely to be related to the environment of the individual, such as family or school experiences. The environmental influence that family members can have upon Maths Anxiety will be explored further in Chapter 5.

> Malanchini et al. (2017) investigated spatial anxiety, Maths Anxiety and General Anxiety in the UK. From a sample of 1,464 twin pairs, aged between 19 and 21 years, they found that all anxieties, including Maths Anxiety, were 30-41% heritable. The remaining percentage is likely to be due to environmental factors.

Genetics may predispose some people to be more likely to experience Maths Anxiety, but this is very hard to measure and determine. On top of this, it's not clear which aspects of genetics make it more likely for someone to be maths anxious. It is important not to ignore the suggestion that a notable portion of the differences in Maths Anxiety may be heritable, and this should be kept in mind when exploring Maths Anxiety. Arguments about the role of genetics can appear quite simplistic, with conclusions being drawn that suggest anything unaccounted for by genetics is down to the environment. However, the reality is likely to be more complex than this, including a multitude of interactions between genetic predispositions and environmental factors, plus many changes over time.

Children Are Less Likely to Experience Maths Anxiety Than Adults

Based on our experience with young students, we do not fully agree with this statement. But then, why do some studies find this to be the case? For example, some researchers suggest that adults are more likely to be maths anxious than children because they have been more exposed to negative experiences related to Maths. This could be negative experiences when learning Maths, such as being embarrassed in a Maths lesson in front of peers, or related to potentially stressful life events, such as a GCSE Maths exam, or becoming financially independent. In comparison, children have not had as many experiences and are therefore less likely to experience Maths Anxiety because of this.

> The Maths Anxiety Trust (2018) commissioned a survey to measure Maths Anxiety in 2,000 students, aged between 11 and 18 years old. The findings suggest that anxiety was at its highest between ages 16 and 17. This may coincide with the time frame of GCSEs and important life choices, such as attending sixth form, college, or joining the workforce. These stressful events in a young person's life could potentially contribute to the experience of Maths Anxiety, leading to higher reports of prevalence at this age.

One explanation for why some researchers identify lower prevalence in children compared to adults is that children may be less able to articulate their feelings clearly or succinctly, meaning that their experiences and emotions, such as Maths Anxiety, are not always understood.

This does not mean that they are not maths anxious. Instead, this may suggest that they are not able to explore and explain their reactions to using and learning Maths as coherently as adults can.

Perhaps, a different way to look at this is that Maths Anxiety can develop at any time, but it may be more likely to increase with age due to having more exposure to mathematical situations, inside and outside the classroom. Counter to this, however, is the argument that more exposure brings more familiarity and practice, which could in turn improve confidence. Again, this way of thinking might be too simplistic. For instance, it might not be the frequency of Maths experiences that is important, but rather the individual events themselves and the lasting effect they could have. Either way, the balance of how much Maths Anxiety research has taken place seems to have fallen in favour of adults. This is possibly due to the ease of access to adult samples. It could also be that there are assumptions that younger people are less likely to experience Maths Anxiety. However, having this preconception may limit the recognition of Maths Anxiety in children, as educators or researchers may not have the awareness needed to successfully support them.

> Hembree (1990) conducted a meta-analysis, which compared different studies of Maths Anxiety. Out of 122 studies included, only 7 of these involved participants less than 16 year olds. None of the participants were younger than 10 years old.

Interestingly, more and more studies are beginning to evidence how Maths Anxiety is likely to develop during childhood, even as young as four years old. Maths Anxiety is mostly reported in Key Stage Two though, when students are aged between 7 and 11 years old. While observable Maths Anxiety is suggested to be less prominent in children, compared to adults, this again may not always be true. Maths Anxiety may not be anticipated by teachers and therefore may go unnoticed. Therefore, it may not be the prevalence of Maths Anxiety that is lesser in children, but perhaps the awareness that they can experience it, or the way it is observed or measured, that accounts for the differences in adults and children.

> Petronzi et al. (2017) investigated the Maths experiences of children in Reception, Year 1 and Year 2 classes across three schools in the UK. Those children who were apprehensive and relatively anxious had a fear of failure and punishment. Discussions with the children led to the suggestion that teachers were seen as a figure of punishment, and that Maths was seen as competitive and hierarchical. This suggests that negative emotions towards Maths can develop at a young age.

It is our belief that individuals of any age may experience Maths Anxiety, and that it is important that, when educators and researchers discuss Maths Anxiety broadly, it should be in reference to all ages, not just adults. By doing so, we are bringing children's experiences to the forefront alongside those of adults and adolescents, and helping them with their experience of Maths Anxiety too, before it becomes increasingly detrimental or difficult to address with age.

Women Are More Maths Anxious Than Men

Since the 1970s, this has been one of the most discussed topics in the field of Maths Anxiety. At this time, researchers wanted to find out why females had poorer mathematical performance compared to males in Western countries, such as the US and the UK. Maths Anxiety and general negative emotions and attitudes towards Maths were suggested to be possible reasons behind this. Nowadays, the performance gap that was once a cause for concern has been shown to be minimal, or even non-existent. However, is there a gender gap in Maths Anxiety? (As a brief note, the literature on this topic often uses the terms gender and sex interchangeably. Gender is the most frequently used term, so that is what we have chosen to opt for here).

> Devine et al. (2012) measured performance and Maths Anxiety in 433 secondary school students in the UK and found that boys and girls didn't differ in their mathematical performance, but girls reported higher Maths Anxiety than boys.

Many studies claim that females are more maths anxious than males, often with statistics to back this up. For example, the nationwide survey carried out by King's College London and National Numeracy in 2023 found out that, from a pool of approximately 2,000 adults aged between 16 and 75 years old, women were more than twice as anxious as men about using Maths; 13% of men said they experienced this anxiety, compared to 29% of women. Those numbers are pretty powerful.

This has also been found around the globe, with organisations and researchers indicating that women are more susceptible to experiencing Maths Anxiety than men. Interestingly, in some countries such as Jordan, the United Arab Emirates, and Qatar, boys were more anxious than girls. Does this illustrate the role of culture in the relationship between gender and Maths Anxiety? It's an interesting thought, but we need more research to help us explore that further.

> OECD (2013) reported on gender differences in Maths Anxiety observed in the PISA 2012 survey. In most participating countries, girls reported stronger feelings of Maths Anxiety than boys. Gender differences were the most distinct in Denmark, Finland, and Liechtenstein, with a minimum difference of 20% between genders.

Some researchers have tried to explain these gender differences in Maths Anxiety. For example, it has been suggested that females are more willing to admit their feelings of anxiety compared to males, which could mean that observed differences are a product of the way in which Maths Anxiety is measured, rather than reflecting the reality of how men and women feel. However, there are three key points to note here. Firstly, as discussed above, there is a fairly clear pattern of gender differences across countries, spanning a range of geographical regions, cultures, and economies. Secondly, gender differences have been reported in studies for several decades. Finally, many studies now incorporate anonymisation into surveys, especially with the increased use of online, digital surveys. Anonymous responding means that people are more likely to report how they truly feel, without fear of judgement. This leads us to suggest that, on the whole, there is likely to be a difference in Maths Anxiety between boys and girls, and men and women.

Having said that, we suggest some cautionary notes. A consistent pattern does not necessarily infer a big difference. In fact, where gender differences in Maths Anxiety exist, they are often fairly small. Also, as with many variables, there is much overlap, such as plenty of men being more maths anxious than women, and vice versa. This highlights that it's not a straightforward generalisation that fits everyone, and making this assumption can limit our understanding of the experience. Simply comparing Maths Anxiety between groups in this way does not tell us anything about the way in which Maths Anxiety might operate for men and women separately, either. We should also be aware of the differences in Maths Anxiety itself, as it has been reported that females are more likely to experience trait Maths Anxiety in particular, compared to men. It has also been argued that the overlap between Maths Anxiety and General Anxiety could explain gender differences in Maths Anxiety, based on findings that women are generally more anxious than men.

> Goetz et al. (2013) examined gender differences in Maths Anxiety. They reported that females had higher trait Maths Anxiety, which is habitual, compared to males, but there were no gender differences in state Maths Anxiety, which is momentary.

Conversely, other researchers claim that there is no gender difference when it comes to Maths Anxiety. A lack of gender differences have been found by studies in primary schools, secondary schools, higher education, and adults, so it is clear that we cannot claim, with absolute confidence, that women are consistently more likely to be maths anxious than men.

> Maths Anxiety Trust (2018) carried out a nationwide survey, which found that, out of 2,000 students, aged between 11 and 18 years old, there was little difference between females and males in reported levels of Maths Anxiety.

But what about changes over time, and the potential influence of others? We will touch on the importance of teachers a little later, but some researchers have considered the relevance of stereotype threat. In fact, this is often brought up in discussions about Maths Anxiety and gender differences.

Stereotype threat is a phenomenon that sees some people unconsciously fear confirming a negative stereotype about their performance in a particular domain. For instance, this could be regarding expected Maths performance associated with being female. Interestingly, some research findings point to no existence of a stereotype threat effect when it comes to gender and performance in Maths, with the suggestion that there is a publication bias towards publishing data that confirms the existence of stereotype threat. However, studies that combine the notion of stereotype threat, Maths Anxiety, and Maths performance are needed.

While there may be some evidence for a gender difference, the inconsistency in research pulls the significance and magnitude of this difference into question. Does this mean that genetics have a role to play in Maths Anxiety in relation to gender? Or is it the socialisation of different genders that can lead to increased susceptibility to Maths Anxiety? No one can say with certainty, and delving deep into the debate surrounding the relationship between gender and Maths Anxiety can bring up more questions than we began with.

> Keshavarzi and Ahmadi (2013) measured Maths Anxiety in 834 secondary school students, consisting of 366 male participants and 468 female participants. Their findings indicated no statistically significant difference between male and female levels of Maths Anxiety.

There certainly seems to be a general pattern, whereby girls and women report higher Maths Anxiety than boys and men. But, we need to bear in mind that these are very large and heterogeneous groups of individuals, and any one study typically captures gender within a specific context, whether that is an age group or phase of education, a geographical region, or a particular socio-economic group. We should always be mindful of the relevance of gender, particularly some of the expectations and assumptions that might surround it when it comes to Maths. However, too much of an emphasis can also further assumptions, such as boys being less likely to experience Maths Anxiety, resulting in a risk of missing boys who need support.

So… Who Has It?

- Maths Anxiety is a widespread problem.
- Prevalence statistics vary widely, depending on the sample, measurement tools, and interpretation of findings.
- Different countries report different levels of Maths Anxiety.
- Background variables such as culture and socio-economic status might contribute to the development of Maths Anxiety.
- Genetics may result in at least some predisposition towards being Maths Anxious.
- People of all ages can experience Maths Anxiety.
- There may be a gender difference Maths Anxiety, but evidence is not consistent.
- Anyone can be maths anxious – don't discount anyone if they don't fit the mould.

Summarising "who has it?" is quite tricky. With varied prevalence, informed by where and how the research was carried out, as well as who was part of the sample, it is challenging to state with conviction the statistical proportion of the population that is maths anxious. On top of this, there may be a myriad of things that affect reported prevalence, such as an individual's country and culture, socio-economic status, biological predisposition, age, or gender.

Even though possible trends in the data may be helpful when trying to determine who might experience Maths Anxiety, we can't make assumptions that some people are definitely more prone to be maths anxious than others. This could lead to someone who is maths anxious not getting the support they need to improve their experience and relationship with Maths, because they didn't fit the expectations.

We suggest to keep your mind open – Maths Anxiety can truly be experienced by anyone, but it also seems that some people might have a greater likelihood of experiencing it compared to others; it really is quite varied.

References

Ashcraft, M. H. and Moore, A. W. (2009) 'Mathematics Anxiety and the Affective Drop in Performance', *Journal of Psychoeducational Assessment,* 27, pp. 197–205. Available at: http://dx.doi.org/10.1177/0734282908330580.

Ashcraft, M. H. and Ridley, K. S. (2005) 'Math Anxiety and Its Cognitive Consequences: A Tutorial Review', in Campbell, J. I. D. (ed.) *Handbook of Mathematical Cognition.* New York: Psychology Press, pp. 315–327.

Betz, N. (1978) 'Prevalence, Distribution, and Correlates of Math Anxiety in College Students', *Journal of Counseling Psychology,* 25, pp. 441–548. Available at: https://doi.org/10.1037/0022-0167.25.5.441.

Burns, M. (1998) Math: Facing an American Phobia. California: Math Solutions Publications.

Chinn, S. (2009) 'Mathematics Anxiety in Secondary Students in England', *Dyslexia,* 15(1), pp. 61–68. Available at: https://doi.org/10.1002/dys.381.

Devine, A., Fawcett, K., Szűcs, D. and Dowker, A. (2012) 'Gender Differences in Mathematics Anxiety and the Relation to Mathematics Performance While Controlling for Test Anxiety', *Behavioral and Brain Functions,* 8, pp. 1–9. Available at: https://doi.org/10.1186/1744-9081-8-33.

Eccius-Wellmann, C., Lara-Barragán, A. G., Martschink, B. and Freitag, S. (2017) 'A Comparison of Mathematics Anxiety Profiles Between Mexican and German Students', *Revista Iberoamericana de Educación Superior,* 8(23), pp. 69–83.

Fan, X., Hambleton, R. K. and Zhang, M. (2019) 'Profiles of Mathematics Anxiety Among 15-Year-Old Students: A Cross-Cultural Study Using Multi-Group Latent Profile Analysis', *Frontiers in Psychology,* 10, pp. 1–9. Available at: https://doi.org/10.3389/fpsyg.2019.01217.

Goetz, T., Bieg, M., Lüdtke, O., Pekrun, R. and Hall, N. C. (2013) 'Do Girls Really Experience More Anxiety in Mathematics?', *Psychological Science,* 24(10), pp. 2079–2087. Available at: https://doi.org/10.1177/0956797613486989.

Gough, M. F. (1954) 'Mathemaphobia: Causes and Treatments', *The Clearing House,* 28(5), pp. 290–294. Available at: https://dx.doi.org/10.4236/psych.2013.46A2005.

Guzman, B., Rodriguez, C. and Ferreira, R. A. (2021) 'Longitudinal Performance in Basic Numerical Skills Mediates the Relationship Between Socio-Economic Status and Mathematics Anxiety: Evidence from Chile', *Frontiers in Psychology,* 11, pp. 1–11. Available at: https://doi.org/10.3389/fpsyg.2020.611395.

Hart, S. A. and Ganley, C. M. (2019) 'The Nature of Math Anxiety in Adults: Prevalence and Correlates', *Journal of Numerical Cognition,* 5(2), pp. 122–139. Available at: https://doi.org/10.5964%2Fjnc.v5i2.195.

Hembree, R. (1990) 'The Nature, Effects, and Relief of Mathematics Anxiety', *Journal for Research in Mathematics Education,* 21(1), pp. 33–46. Available at: https://doi.org/10.2307/749455.

Johnston-Wilder, S., Brindley, J., and Dent, P. (2014) Technical Report: A Survey of Mathematics Anxiety and Mathematical Resilience Amongst Existing Apprentices. Coventry: University of Warwick.

Jones, W. G. (2001) 'Applying Psychology to the Teaching of Basic Math: A Case Study', *Inquiry: A Journal of Medical Care Organization,* 6(2), pp. 60–65.

Karim, A. H., Hajar, M. S. N. and Tunku Ahmad, T. B. (2023) 'Prevalence of Math Anxiety Among Upper Secondary Students at a Private School in Suburban Kuala Lumpur', *Jurnal Pendidikan Sains Dan Matematik Malaysia,* 13(1), pp. 52–63..

Keshavarzi, A. and Ahmadi, S. (2013) 'A Comparison of Mathematics Anxiety Among Students by Gender', *Procedia-Social and Behavioral Sciences,* 83, pp. 542–546. Available at: https://doi.org/10.1016/j.sbspro.2013.06.103.

Lane, C. (2017) 'Student's Images of Mathematics: The Role of Parent's Occupation', in Dooley, T. and Gueudet, G. (eds.) Proceedings of the Tenth Congress of the European Society for Research in Mathematics Education (CERME10). Dublin: ERME, pp. 1130–1337.

Malanchini, M., Rimfeld, K., Shakeshaft, N. G., Rodic, M., Schofield, K., Selzam, S., Dale, P. S., Petrill, S. A. and Kovas, Y. (2017) 'The Genetic and Environmental Aetiology of Spatial, Mathematics and General Anxiety', Scientific Reports, 7(1), p. 42218. Available at: https://doi.org/10.1038%2Fsrep42218.

Maths Anxiety Trust (2018) Maths Anxiety Summit 2018 Report Final. Available at: https://mathsanxietytrust.com/Maths-Anxiety-Summit-2018-Report-Final-2018-08-29-1 (Accessed: 7 April 2024).

National Numeracy (2023) A Third of Adults Are Nervous About Numbers. Available at: https://www.nationalnumeracy.org.uk/news/third-adults-are-nervous-about-numbers (Accessed: 29 April 2024).

OECD (2013) PISA 2012 Results: Ready to Learn: Students' Engagement, Drive and Self-Beliefs (Volume III). Paris: OECD Publishing. Available at: https://dx.doi.org/10.1787/9789264201170-en.

OECD (2015) Does Math Make You Anxious? PISA in Focus, 48. Paris: OECD Publishing. Available at: https://doi.org/10.1787/5js6b2579tnx-en.

OECD (2023) *PISA 2022 Results (Volume I): The State of Learning and Equity in Education*. Paris: OECD Publishing. Available at: https://doi.org/10.1787/53f23881-en.

Perry, A. B. (2004) 'Decreasing Math Anxiety in College Students', *College Student Journal*, 38(2), pp. 321–324.

Petronzi, D., Staples, P., Sheffield, D., Hunt, T. and Fitton-Wilde, S. (2017) 'Numeracy Apprehension in Young Children: Insights from Children Aged 4–7 Years and Primary Care Providers', *Psychology and Education*, 54(1), pp. 1–23.

Richardson, F. C. and Suinn, R.M. (1972) 'The Mathematics Anxiety Rating Scale: Psychometric Data', *Journal of Counseling Psychology*, 19(6), pp. 551–554. Available at: https://doi.org/10.1037/h0033456.

Short, D. and McLean, J. (2023) Assessment of Maths Anxiety in Early Schooling: Emergence, Stability and SES Differences. Loughborough: Mathematical Cognition and Learning Society.

Thilmany, J. (2009) 'Math Anxiety', Mechanical Engineering-CIME, 131(6), pp. 11–12.

Wang, Z., Hart, S. A., Kovas, Y., Lukowski, S., Soden, B., Thompson, L. A., Plomin, R., McLoughlin, G., Bartlett, C. W., Lyons, I. M. and Petrill, S. A. (2014) 'Who Is Afraid of Math? Two Sources of Genetic Variance for Mathematical Anxiety', *Journal of Child Psychology and Psychiatry*, 55(9), pp. 1056–1064. Available at: https://doi.org/10.1111/jcpp.12224.

Yanuarto, W. N. (2016) 'Teachers Awareness of Students' Anxiety in Math Classroom: Teachers' Treatment VS Students' Anxiety', *Journal of Education and Learning (EduLearn)*, 10(3), pp. 235–243. Available at: https://dx.doi.org/10.11591/edulearn.v10i3.3808.

3 Can I Measure It?

Maths Anxiety seems to be quite widespread, but frustratingly, we don't actually know for certain how many people experience it. Sometimes Maths Anxiety can be confused with negative attitudes towards Maths, or other emotions, so how can we actually tell if someone is experiencing Maths Anxiety? If we can find this out, can we also tell if someone is mildly or extremely maths anxious, or how maths anxious they are compared to others? Is Maths Anxiety even a measurable experience?

Researchers have worked hard to develop tools and techniques to measure Maths Anxiety, which helps us to answer these questions. Being able to measure Maths Anxiety has definitely helped us gain much-needed information, and it has provided data that we can work with, and from. For example, measuring Maths Anxiety might give a rough indication of prevalence in a sample of participants, suggest possible relationships with other variables, or indicate the make-up of Maths Anxiety.

The hard part is that there are so many options for us to choose from. There are quite a few methods available to measure Maths Anxiety, or at least certain responses that are thought to be associated with Maths Anxiety. These include physiological measures, measures of brain activity, self-reporting measures like questionnaires, or qualitative methods such as observations or interviews. This can sometimes lead to confusion or conflicting evidence, so it can be tricky to know where to begin when we want to measure Maths Anxiety.

> Atkinson (1988) reflected on the history of Maths Anxiety research and broke this down into three distinct areas, spanning 30 years between the 1950s and 1980s:
>
> 1 The first period of measuring and reporting Maths Anxiety was based on authors' opinions. No actual standardised measures were used, as claims were based on anecdotal evidence or common sense at the time, looking more at "emotional blocks" to learning Maths, rather than specifically categorising Maths Anxiety. This was during the time that Maths Anxiety was beginning to be defined, around the 1950s.
> 2 Once the definitions of Maths Anxiety became clearer, the second period emerged where researchers tried to assess a range of attitudes towards Maths, to get a more defined picture of Maths Anxiety. These surveys included aspects such as

> confidence, enjoyment, attitudes towards Maths, misconceptions, and state or trait anxiety.
> 3 The third period of research saw questionnaires being continually developed and refined, so that they could best measure Maths Anxiety. They were put to the statistical test and were held more accountable than ever before.

Physiological Measures

> Demedts et al. (2023) conducted a study of physiological data and Maths Anxiety. Their findings point to physiological measures offering a possible contribution to understanding the construct of Maths Anxiety but warned about being too optimistic about this measurement method.

One way in which Maths Anxiety has been measured is physiologically. For example, when faced with a mathematical task or situation involving Maths, a maths anxious person may have:

- Heart rate variability (sometimes simply an increased heart rate);
- Raised skin temperature;
- Increased levels of cortisol (stress hormone);
- Changes in breathing patterns;
- Rapid eye movements and pupil dilation;
- Increased perspiration (sweat);
- Heightened galvanic skin response (where the skin becomes a better conductor of electricity)

> Mattarella-Micke et al. (2011) measured the secretion of cortisol in participants just before they were presented with challenging mathematical problems. Those with high Maths Anxiety and a high working memory capacity showed a negative relationship between performance on the mathematical problems and cortisol secretion. That is, the higher their cortisol secretion, indicating high stress and anxiety levels, the worse their performance was. Interestingly, this wasn't found for individuals with low working memory.

These bodily responses are similar to General Anxiety Disorder and other forms of anxiety, so they appear to be a reliable way to measure someone's Maths Anxiety. But is it a realistic approach for a typical educational context? Physiological measures of Maths Anxiety aren't particularly appropriate for the classroom, with various pieces of equipment being hooked up to children during a Maths lesson (that would definitely be a health and safety issue). These methods may be good to use in a laboratory, but not as practical in the actual place where the majority of learning and using Maths happens the most – the classroom. Being aware of the possible physiological changes associated with Maths Anxiety, however, may prompt you to notice behaviours related to these responses, such as appearing flush in the face, or wiping sweat off the forehead.

> Hunt, Bhardwa, and Sheffield (2017) assessed changes in heart rate and blood pressure in 77 primary school children when they were presented with Maths problems of increasing difficulty. Children's systolic blood pressure was found to increase when presented with the most difficult problems. This change was positively correlated with children's self-reported Maths Anxiety, suggesting that a very real stress response can be observed amongst those who feel more maths anxious. Moreover, these findings suggest a need to be aware of subtle features of Maths that might result in a particular response from someone high in Maths Anxiety, such as how difficult problems are.

While useful, the success of these physiological measures may depend on everyone having a similar bodily reaction to anxiety or stress. For example, when you are feeling anxious your heart rate might increase, but the extent of the change might be different to others'. This individual variation in physiological reactivity may make it difficult to make consistent links with Maths Anxiety. That said, it does seem that heart rate (and heart rate variability) or cortisol secretion may be the most reliable measures in terms of consistency, and they are also amongst the most regularly used physiological measures for assessing stress responses to Maths. In particular, psychophysiological reactivity recorded at the level of the individual student offers a way of assessing changes that are specific to that person. Certain measures, such as eye-tracking, can also provide evidence of a direct link with behavioural responses. For instance, we might ask whether those who are high in Maths Anxiety simply look, or don't look, at a Maths task differently to those who are low in Maths Anxiety.

> Hunt, Sheffield, and Clark-Carter (2014) found that Maths Anxiety was associated with a range of eye movements (fixations, dwell-time, and saccades) during a Maths task.

Gradually, researchers have devised more complex experimental designs in line with advances in technology. For example, more recent eye-tracking equipment is quicker to use and is less intrusive. This enables closer investigation of potential cognitive processes associated with Maths Anxiety, usually alongside measures of Maths performance.

Brain Activity

Another way to measure Maths Anxiety is to look at brain activity, to see whether there are changes in particular areas of the brain, or if certain regions in the brain are activated in situations involving Maths. This is usually done through functional magnetic resonance imaging (fMRI), which has been recently used with child participants as well.

When testing for any form of anxiety, evidence has shown how stress can impact regions of the brain's prefrontal cortex, which is involved in attention and decision-making, personality, and social behaviour, and regulating the expression of fear. In particular, those with anxiety may have altered brain activation in the right dorsolateral prefrontal cortex and left inferior frontal sulcus and increased deactivation of the rostral-ventral anterior cingulate cortex. The amygdala, famously known for being in charge of our "fight or flight" response, has also been associated with Maths Anxiety. If someone is maths anxious, they may be more likely to have increased activity in their amygdala.

> Young, Wu, and Menon (2012) carried out an fMRI study with 7-9 year olds and reported that Maths Anxiety was associated with increased activity in the right amygdala regions (known to be important for the processing of negative emotions), and reduced activity in the posterior parietal and dorsolateral prefrontal cortex regions (which are involved with mathematical reasoning). They also found high Maths Anxiety to be associated with greater connectivity between the amygdala and the ventromedial prefrontal cortex regions, which was unrelated to general anxiety, attainment, working memory, or reading performance. Measuring Maths anxiety in this manner highlighted its link to specific brain activity.

Similar to physiological measures, tracking and measuring brain activity is definitely not something we can do easily in the classroom. It is useful though, as researchers doing this on our behalf have scientifically shown that Maths Anxiety can be identified and assessed through tracking changes in certain regions of the brain, indicating its presence, and potentially, its growth.

An additional issue is that a high level of understanding and training is needed to carry out neuroimaging such as fMRIs, and to understand the results. It is rather specific and although it can demonstrate the presence and impact of Maths Anxiety, it is not easy to interpret. Also, it is usually the case that data from several people are analysed together in a way that patterns are identified, rather than taking measurements of brain activity from a single person and saying *"We can see Maths Anxiety in that person's brain!"*.

> Pletzer et al. (2015) aimed to compare fMRI results between mathematically anxious and non-mathematically anxious participants with similar Maths performance. When solving non-numerical tasks, their brain activation patterns were the same. When they were completing numerical tasks, all participants had similar activation in the intraparietal sulcus, which is involved in number processing, but mathematically anxious participants had more activity in the default mode network, which is primarily made up of the prefrontal cortex, posterior cingulate cortex, and angular gyrus. This suggests that maths anxious individuals may have impaired processing efficiency compared to those who are less anxious.

Similar to fMRI, a handful of studies have utilised electroencephalography to study electrical brain activity in association with Maths Anxiety. An electroencephalogram (EEG) enables researchers to measure quite specific electrical activity in response to the presentation of stimuli, such as Maths problems, and at specific times. This can be useful when it comes to identifying areas of the brain involved in mathematical cognition, including how they might vary according to someone's Maths Anxiety. Other than the typical practical constraints associated with this approach, EEG poorly measures neural activity that occurs below the upper layers of the brain (the cortex), which does limit how much we can gain from its use in the context of Maths Anxiety.

What can we actually do as educators? How can we measure Maths Anxiety if we don't have a laboratory, equipment, or specialist training? Perhaps more importantly, how can we measure Maths Anxiety in a more direct way? After all, physiological and brain activity measurements on their own will probably not cut it.

Questionnaires

Self-reporting tools, such as questionnaires, are definitely one way we can measure Maths Anxiety in the classroom. This is seen as a quantitative method for measuring Maths Anxiety, meaning that the information you collect from them are based on numbers. While some questionnaires can ask open questions and require written responses, Maths Anxiety questionnaires (MAQs) often ask people to circle their responses, in the form of a number on an ordinal scale.

A typical approach would be to present a series of statements pertaining to Maths situations, then ask people to select how anxious they would feel in each of those situations. Responses are usually on a scale of, say, 1 "not at all anxious" to 5 "extremely anxious," or something along those lines. To avoid confusion, we should also point out the varied language that is sometimes used. The term "questionnaire" is fairly broad and tends to refer to the use of a series of questions that in some way measure Maths Anxiety. The term "self-report measure" indicates that the questionnaire itself is given to a person for them to complete, without intervention or assistance from anyone else. The term "scale" is typically reserved for self-report measures that are purely quantitative in nature and contain items/ questions/ statements that all relate to the measurement of a particular thing, such as Maths Anxiety. Here, we might vary the terms we use, and we often refer to "questionnaires" for brevity, but it is useful to know what we mean.

There have been lots of questionnaires developed and tested over the years in order to measure Maths Anxiety. They can suggest the presence of Maths Anxiety, measure how maths anxious someone is, and understand more about experiences numerically. They give you tangible data, which you might want if you are doing action-based research in school for example, or providing a snapshot of what Maths Anxiety is like at your school or in a certain class or group.

While we will be exploring particular MAQs in detail, Table 3.1 displays a list of psychological variables that are often studied alongside Maths Anxiety – some more than others. The list is by no means exhaustive, and variables are listed alphabetically rather than indicating importance or priority. Alongside key psychological constructs are some examples of validated self-report measures, including subscales where relevant.

When discussing Maths Anxiety at a conceptual level, the term "factors" is often used. This is a statistical term that represents the way in which scores on individual questions, often termed "items," might indicate that an overall construct (such as Maths Anxiety in this case) is composed of multiple underlying sub-constructs. The term "subscales," on the other hand, refers to the specific parts of an overall scale that enable us to measure those factors. If you are looking at conducting your own research into Maths Anxiety, Table 3.1 could be a useful reference point to help you get started.

Table 3.1 Psychological Variables and Self-report Measures Related to Maths Anxiety

Variable	Example Measure and Subscales	Description
Appraisal of Previous Mathematics Experiences	Hunt and Maloney (2022) Appraisals of Previous Mathematics Experiences Scale **Subscales:** Appraisal of everyday Maths experiences Appraisal of Maths evaluation experiences	This construct relates to how someone appraises previous Maths experiences, i.e. the extent to which they view their experiences negatively or positively.
Approach – Avoidance	Elliot and Murayama (2008) Achievement Goal Questionnaire –Revised **Subscales:** Mastery-approach Mastery-avoidance Performance-approach Performance-avoidance	This represents how a student's motivations might be based on a combination of wanting to aim for mastery (of Maths learning/ understanding) and high performance, but also the avoidance of not achieving mastery or high performance.
Extrinsic / Instrumental Motivation	OECD (2014) Programme for International Student Assessment	This construct relates to how motivated a student is to engage with Maths for the purpose of what success in Maths can bring, i.e. beyond the Maths itself.
Fear of Failure	Conroy (2001) Performance Failure Appraisal Inventory (short form)	This construct reflects the fear that might be experienced at the prospect of, actual, or perceived failure in the context of Maths learning and performance.
Growth Mindset	Kooken et al. (2016) Mathematical Resilience Scale	This construct refers to the extent to which a student believes their knowledge and understanding of Maths can improve. There appears to be some conceptual overlap with the concept of hopelessness (as measured by the AEQ), but hopelessness is a less-studied construct in the context of Maths.
Intrinsic Motivation	Adapted from Vallerand et al. (1992) Academic Motivation Scale	This construct relates to how motivated a student is to engage with Maths based on thoughts and feelings about Maths itself, rather than any perceived outcomes associated with Maths success. There is likely to be some overlap with Maths enjoyment, but intrinsic motivation is a little broader, e.g. encompassing perceived satisfaction with, and beauty in, Maths.

Table 3.1 Psychological Variables and Self-report Measures Related to Maths Anxiety (continued)

Variable	Example Measure and Subscales	Description
Intrusive Thoughts	Freeston et al. (1991) Cognitive Intrusions Questionnaire (CIQ)	This construct represents unwanted, negative thoughts that might experienced by students, usually on an instantaneous basis and typically experienced during particular Maths situations, e.g. during heightened arousal in high-stakes testing situations or during Maths learning that detrimentally impacts a student's Maths self-efficacy, confidence, motivation, self-concept, etc. The CIQ contains several – and different – questions pertaining to intrusive thoughts. These are related to the nature of the thoughts, a person's perceptions of how they deal with thoughts, and how they feel the thoughts have impacted them.
Maths Norms	OECD (2014) Programme for International Student Assessment	This construct refers to a student's perceptions of others in the context of Maths. Typically, this would be in the context of peers and parents, e.g. beliefs about fellow students' work ethic associated with maths learning, or beliefs about parental expectations about Maths attainment. For an adult population, it is less appropriate to refer to parents. As such, it might be more appropriate to focus on friends, or work colleagues.
Maths Self-concept	OECD (2014) Programme for International Student Assessment	This construct refers to how a student sees themselves in the context of Maths. It is reasonably broad, but still within the context of Maths.
Maths Self-efficacy	OECD (2014) Programme for International Student Assessment	This construct relates to a learner's confidence in their ability to successfully perform a specific maths task, i.e. ahead of performing the task.

Table 3.1 Psychological Variables and Self-report Measures Related to Maths Anxiety (continued)

Variable	Example Measure and Subscales	Description
Other Maths Emotions	Pekrun et al. (2011) Achievement Emotions Questionnaire (AEQ) **Subscales:** Pride Anger Boredom Shame Enjoyment Hopelessness	The AEQ is generally used to assess students' emotions in the context of learning, but it has been used in the specific context of Maths learning. This includes a range of relevant emotions.
Perceived Importance of Struggle	Kooken et al. (2016) Mathematical Resilience	This construct represents the extent to which a student believes that struggle is an important component of successful Maths learning.
Perceived Usefulness of Maths	Lim and Chapman (2013) Attitudes Towards Mathematics Inventory (short form)	This construct is a specific attitude towards Maths, related to how useful a student believes maths to be in the context of its applications. This is typically conceptualised in reference to broad contexts such as work or everyday life. This is closely related to instrumental motivation.

Using a questionnaire will usually provide you with numerical data, but potentially richer information as well. For example, you might be able to tell what situations make an individual nervous, or see whether it's the apprehension of tasks or the tasks themselves which create feelings of anxiety. Using questionnaires to compare results within your class, or across classes or cohorts, will also provide context to your findings. For example, comparing individual results to that of the class, or the class's results to another class, will perhaps demonstrate where the experience, or experiences, "sit" in relation to others. The approach also has the advantage of enabling us to measure changes over time. Having this information will allow you to understand the experience of Maths Anxiety that you are exploring in a useful, purposeful way.

> Dreger and Aiken (1957) were the first to attempt to measure Maths Anxiety through a questionnaire. They replaced three Maths questions which had low validity in the existing Taylor Manifest Anxiety Scale (Taylor, 1953) with questions about anxiety when working with numbers. They called it the Numerical Anxiety Scale.

Throughout the development of various MAQs, an important discussion between academics in the field of Maths Anxiety has arisen. Is Maths Anxiety a unidimensional construct, or multidimensional? That is, is Maths Anxiety a single "thing," or does it consist of many "things" that require separate measurement?

Traditionally, Maths Anxiety has been utilised as a unidimensional construct. However, as you will gather throughout this book, Maths Anxiety is a multidimensional experience, so questionnaires are often inclusive of more than one factor, or "dimension." These can vary slightly according to the questionnaire used. For example, one might measure three factors, such as Calculation Anxiety, Classroom Anxiety, and Test Anxiety in the context of Maths, while a different questionnaire may measure only two, such as Maths Test Anxiety and Numerical Anxiety. Therefore, it's important that if you are using a questionnaire to measure Maths Anxiety, you need to identify what factors in particular it is measuring, and make sure it's the most suitable for your purpose.

Questionnaires are easy to administer, are often easy to access, and require minimal training and instruction. If you are interested in using a questionnaire to measure Maths Anxiety, hopefully knowing more about their length, possible application to your context, and what they focus on, may help narrow down which is the best for you to use. The questionnaires for both adults and children that we will discuss in this book are generally the most used within existing Maths Anxiety research, but there are many others that you can explore.

The majority of commonly used Maths Anxiety self-report measures were devised in Western, educated, industrialised, rich, democratic (WEIRD) countries. This means that many scales have been created in English and may not be culturally relevant to all countries. In fact, most are designed with the American population in mind. Some of the questions are transferable and applicable in a variety of cultures, but some terminology is very specific. For example, the phrases "pop quiz," "making change," or "dimes and nickels" are reflective of American culture. If these questionnaires are used in different countries but are not adapted, this could mean that some questions are misunderstood or misinterpreted, and therefore may not give the truest picture of Maths Anxiety. This is why many questionnaires are adapted by researchers for use in different countries, so that they are more in line with their language and culture. But, it must be remembered that modifications to scales in this way can potentially impact how reliable and valid they are, so care should be taken not to modify scales too heavily. It is always best to identify a suitable measure that has already been designed and tested with the group of people you have in mind.

If you ask a student directly if they are feeling anxious, they may not be honest in their response. This is usually because they don't feel comfortable, or don't want to share their emotions. When they are asked these types of questions in the form of a written questionnaire, does this make them more likely to be honest? Not necessarily. Many researchers share their frustration with how unreliable questionnaires are, because not everyone is truthful when completing questionnaires. They may be trying to preserve a certain self-image, leading to dishonest answers, or answer in a way in which they think you want. For example, if you are their teacher, they may try to impress you and therefore circle "not anxious at all" for everything. Do bear this in mind when using questionnaires, as they are unlikely to be completely valid and reliable. Therefore, methods to ensure anonymity when completing such measures should be considered where possible.

It is also important to reflect on how different professionals might use MAQs. Researchers will often be interested in assessing group differences, or testing the ways in which Maths Anxiety is associated with a range of psychological and behavioural variables, such as self-belief and attainment. This usually means they are interested in how Maths Anxiety varies in some way within a group, often from an academic perspective. On the other hand, teachers (and various other professionals working in educational settings, such as educational psychologists or specialists in specific learning difficulties) might feel that it is more useful to understand the degree of Maths Anxiety that is reported. This could be at a whole-class or whole-school level, in order to explore the impact of new teaching strategies on Maths Anxiety levels, for example. It could also be that they would like to use questionnaires as a tool for identifying students who might be in need of support, such as those who are particularly maths anxious, or at least more maths anxious than their peers.

This comparison between individuals, while useful, does lead to the question of what actually constitutes a "low" or "high" level of Maths Anxiety. With no universal agreement of what level or score on a questionnaire depicts a "highly maths anxious" individual compared to someone who has "low Maths Anxiety," this can make it difficult to use these terms with certainty. After all, MAQs measure Maths Anxiety on a continuum, but one that has no clear boundaries to support the definitions of "high" or "low." This is something to consider when using questionnaires to measure Maths Anxiety, and how you will reflect on the data you collect.

This also brings to the light the issue of labelling. That is, are there unwanted or unintentional consequences of telling a person that they are "high in Maths Anxiety" or "highly maths anxious," for example? Perhaps, the most problematic language is labelling a student as someone "with" high Maths Anxiety, or as someone who "is" highly maths anxious. While we don't necessarily suggest avoiding such terms completely, we advocate care and caution in their use. To suggest someone "has" something implies a qualitative difference compared to others, that they in some way stand apart from their peers. Identifying a student as "having" Maths Anxiety also implies that others do not. This could create a range of challenges, including students feeling that their Maths Anxiety is not being acknowledged, or that the questionnaire score in some way undermines how they feel. For students with a classification of having Maths Anxiety, or a high level of Maths Anxiety, there is also a risk of this reinforcing already-low mathematical self-belief. Conversely, a high Maths Anxiety score on a questionnaire could be very reaffirming, as it might provide them with an external indicator of how they are feeling inside. After all, self-report measures of psychological constructs are designed to reflect how individuals actually feel.

This leads us to the biggest question of all when it comes to self-reporting measures of Maths Anxiety: how do they benefit us in the classroom? For many students, and perhaps even teachers, they might not have previously engaged in the completion of such things. One of the most common experiences reported to us by teachers is that questionnaires provide them, and their students, with a tool to explore and tackle Maths Anxiety. They can offer an alternative way to reflect on, think about, and discuss feelings about Maths, rather than simply relying on talking. Therefore, questionnaires may be hugely beneficial not only in understanding Maths Anxiety, but also in taking action against it. We will cover this in more depth in Chapter 6.

Adult and Adolescent Maths Anxiety Questionnaires

The majority of questionnaires that have been developed are designed for use with adults or adolescents, rather than children. These are useful for a variety of situations. Perhaps, you are a teacher in secondary or further education who would like to better understand the experience of maths anxious students, or you may want to explore Maths Anxiety in the adult population, such as the parents of the children you teach, or even fellow teachers.

From reviewing the available literature, the following are the most commonly used questionnaires to measure adult or adolescent Maths Anxiety. This is not to say these are the only questionnaires available, or the only ones we recommend, but they each have their own strengths and specific factors that they measure.

Richardson and Suinn (1972): Maths Anxiety Rating Scale

This is probably the most well-known questionnaire within Maths Anxiety research and it has been tried and tested in various settings. This questionnaire poses 98 questions to assess how anxious someone would be in different mathematical situations. This is done through a Likert scale, where an individual would read the statement, such as "Getting ready to study for a Maths test," and then respond by selecting the best fitting response, such as (1) not anxious at all, (2) a little anxious, (3) anxious, (4) very anxious, or (5) very, very anxious. Using a Likert scale helps determine the "strength" or "level" of someone's experience of Maths Anxiety, and most questionnaires that measure Maths Anxiety use them.

> Capraro, Capraro, and Henson (2001) wanted to explore how reliable the Maths Anxiety Rating Scale (MARS) was. They reviewed 67 research papers where the MARS was used and found that the MARS had strong internal consistency, meaning that the individual items correlate well with each other (they appear to be broadly measuring the same thing). They also found that the MARS had good test-retest reliability, meaning that if someone completed the test twice at different times, they would receive very similar scores.

The MARS has been found to be very reliable and consistent. It is generally accepted that the MARS measures two factors in detail: Mathematics Test Anxiety, which is anxiety related to Maths tests or the experience of being evaluated, and Numerical Anxiety, which refers to anxiety related to the manipulation of numbers, basic Maths skills and using Maths in everyday situations. The original MARS has 98 questions, which would take a large chunk of time to administer correctly, but due to the MARS's success, it has now been made available in several versions of varying lengths and languages. This makes it more accessible to researchers in various contexts.

> Suinn and Winston (2003) developed an abbreviated version of the MARS. Reportedly, this can measure the same dimensions of Maths Anxiety as the MARS, but by using only 30 questions instead. This is much easier to administer, and it takes a third of the time, also demonstrating high internal consistency.

Fennema and Sherman (1976): Fennema-Sherman Mathematics Attitudes Scale

Fennema and Sherman used 108 questions and a 5-point Likert scale to measure different psychological variables associated with Maths Anxiety, with a focus on gender-related differences. This questionnaire evaluates nine specific areas believed to be integral to understanding the experience:

- Attitudes towards Maths and success;
- Maths as a male subject;
- Perception of parental attitudes towards Maths;
- Perception of teacher attitudes towards Maths;
- Confidence in learning Maths;
- Feelings and bodily symptoms related to dread and nervousness;
- Motivation to use Maths, and;
- Beliefs that Maths is useful.

They noted that these nine areas do overlap, but each domain should be reviewed specifically in order to measure and understand Maths Anxiety in detail. Using 108 questions is quite substantial, but some researchers appreciate the level of detail and specificity that is offered through this questionnaire. Nevertheless, much of this scale focuses on what is associated with Maths Anxiety, rather than focusing specifically on Maths Anxiety itself.

> Sachs and Leung (2007) highlighted the need to shorten the Fennema-Sherman Mathematics Attitudes Scale, as the 108 questions require approximately 45 minutes to answer, leading to the possibility of fatigue, which could impact the validity of responses near the end of the questionnaire.

Wigfield and Meece (1988): Mathematics Anxiety Questionnaire (MAQ)

The Math Anxiety Questionnaire (MAQ) was designed to measure six areas of Mathematics Anxiety, related to the two factors of worry and negative affective reactions:

- Dislike;
- Lack of confidence;
- Discomfort;
- Worry, fear and dread;
- Confusion, and;
- Frustration.

The 22 questions are asked using a 7-point Likert Scale, but this may provide too many options for some people and lead to confusion over which point on the scale to pick. Despite this, many researchers find this questionnaire to be reliable and easy to use.

> Kazelskis and Reeves-Kazelskis (1999) tested the MAQ and found that it was effective in measuring the two factors of worry and negative affective reaction across different ages and gender.

Hopko et al. (2003): Abbreviated Maths Anxiety Scale

This is a relatively short questionnaire, using only nine questions and a 5-point Likert scale to measure anxiety related to learning Maths, and anxiety related to being tested in Maths. It has been held in high regard by researchers and has been used to assess Maths Anxiety in adults and younger students. This questionnaire might be quite useful because it takes less than five minutes to administer and appears to be suitable in various cultural contexts. It has been questioned whether nine questions is enough to truly give enough information about someone's experience of Maths Anxiety though, and therefore be able to provide an accurate measurement. Some people also argue that the measure is a little restrictive in that the items mostly relate to a formal educational context, excluding the wider Maths context.

> Cipora et al. (2015) analysed a Polish adaptation of the Abbreviated Maths Anxiety Scale (AMAS). They found the Polish adaptation of the AMAS to have high internal consistency and reliability, suggesting that this version is also an effective measure of Maths Anxiety. They suggested that more research needs to be conducted to identify whether assessment tools, such as the AMAS, are suitable for use in multiple countries, to better understand Maths education globally.

Hunt, Clark-Carter, and Sheffield (2011): Mathematics Anxiety Scale UK

The Mathematics Anxiety Scale UK (MAS-UK) has been reported to be a reliable measure of Maths Anxiety, and easy to administer. Using 23 questions, participants are asked to think about how anxious they would feel in a variety of different situations: not at all; slightly; a fair amount; much, or very much. The questions focus on three factors, which were identified through exploratory factor analysis (EFA): Maths Evaluation Anxiety, Everyday/Social Maths Anxiety, and Maths Observation Anxiety.

> Firouzian et al. (2015) modified some questions in the MAS-UK to be more suitable for use in the context of Iranian Universities. They reported that the scale had very good internal consistency, with Chronbach's alpha of 0.87. This indicates that MAS-UK is a good measure of Maths Anxiety which can be successfully translated for use within other countries.

In the MAS-UK, the Maths Evaluation Anxiety factor includes questions that relate to being evaluated or tested in the context of Maths. This includes situations such as being asked a Maths question by a teacher, or taking a Maths test. The Everyday/ Social Maths Anxiety factor refers to Maths situations outside of the typical education or assessment setting. This includes questions related to typical situations in everyday life where a person is required to engage in mathematical problem-solving in some way, such as calculating the number of days until a person's birthday, or working out the total cost of your shopping. The Maths Observation Anxiety factor includes questions that refer to situations involving Maths which do not necessarily require actual Maths learning, problem-solving, or evaluation. These situations often involve being a passive observer of Maths, such as reading the word "algebra" or listening to someone talk about Maths.

> Hunt, Clark-Carter, and Sheffield (2011) found that those who report a high level of Maths Anxiety are often the individuals who score highly on the Maths Observation Anxiety factor. This means that the mere presence of maths can be enough to make a person feel anxious and, where students are concerned, being around Maths is quite hard to avoid. This highlights the successful use of questionnaires to not only measure Maths Anxiety, but to understand its complexities further.

Hunt et al. (2019): Mathematics Calculation Anxiety Scale

Some researchers have previously suggested that Maths Anxiety can exist in a way that is specific to numerical calculation, and even more abstract Maths. In fact, many people who declare themselves as being maths anxious will say that algebra is one area of Maths that fills them with dread. Building on this, Hunt and colleagues devised a 26-item questionnaire that is specific to anxiety in the context of Maths calculations. This does not take into account broader context, such as other people or the situation in which calculations are performed, but focuses instead on anxiety associated with being asked to perform a range of mathematical calculations based on the GCSE Maths curriculum in England.

> Hunt et al. (2019) analysed data from 160 university students and found 4 factors underlying the Mathematics Calculation Anxiety Scale: Abstract Maths Anxiety; Statistics Probability Anxiety; Statistics Calculation Anxiety, and Numerical Calculation Anxiety.

Child Questionnaires

MAQs have also been developed with children in mind. Most questionnaires for children's experiences of Maths Anxiety are designed to be reflective of the Maths curriculum and situations that are relevant for children, but some are adult questionnaires made "child-friendly."

The appropriateness of using questionnaires to measure Maths Anxiety in children has been questioned, as children may not be able to reflect on their emotions and anxiety as effectively as adults. This leads to the suggestion that children's responses on a questionnaire may not truly reflect their actual experience. Importantly, the questionnaire used also needs to be appropriate for the age range and ability of the child completing it. Giving children a difficult questionnaire in terms of comprehension and content will undoubtedly increase feelings of anxiety, and potentially impact their responses to the questions.

A further consideration is the use of numerical Likert scales. Rating anxiety numerically is quite an abstract concept and while this might be fine for use with adults, with children it can make it quite challenging. This difficulty increases when there are more points on the scale, such as a 5-point Likert scale or a 7-point Likert scale. On top of this, researchers question the suitability of numerical Likert scales, as they may be too far removed from what they are trying to represent. For example, if we said to you, on a scale of 1-5, how happy are you right now? You might be able to whittle it down to two numbers and then pick one of them, but probably not with absolute certainty that that number genuinely reflects your current level of happiness. What do you think a young child would say? They are likely to understand the difference between happy and unhappy but would they be able to place their happiness on a continuum? Similarly, asking children to share how anxious they are by circling a number may not create an accurate representation of their Maths Anxiety, especially as, ironically, they are being asked about their Maths Anxiety while responding using a number.

This may be the reason why researchers have begun to use pictorial Likert scales to measure Maths Anxiety in children. These use pictures instead of numbers when giving students a choice, which are usually emojis when trying to reflect emotions such as anxiety. Children as young as three years old can understand how symbols can represent objects or basic emotions. This makes pictorial Likert scales a good alternative to numerical Likert scales, as they are easier to understand, are more applicable to a wider range of children, and are more representative and comparable to real life. This is only the case though when the pictures used are actually related to what they are trying to represent. For example, emotions shouldn't be represented through weather drawings such as lightning, rain or sunshine; they should be shown through faces, or emojis, that depict various levels of anxiety.

Below are some questionnaires that are regularly used within research to measure Maths Anxiety in children. You may find these questionnaires useful if you are working with students who are not yet teenagers, but make sure that you look carefully at the suggested age range of each questionnaire, as they are not applicable for all ages without further adaptation.

Suinn, Taylor, and Edwards (1988): Maths Anxiety Rating Scale - Elementary

As the MARS is widely used and celebrated by many researchers, it was adapted for use with children. The Maths Anxiety Rating Scale - Elementary (MARS-E) was developed in the late 1980s and was originally tested with 9-12 year olds. It has 26 questions that ask children about their anxiety in situations involving Maths, such as "When counting how much change you should get back after buying something, how nervous do you feel?" This questionnaire uses a 5-point Likert scale, where children need to pick one of the five choices that best represents their level of anxiety. The MARS-E has been shown to be reliable and easy to administer, and it has also been translated successfully into other languages.

> Baloglu and Balgalmis (2010) adapted the MARS-E so that it could be used in Turkey. To do this, the questions were first translated, and then it was tested for validity. They found that the adapted version had good content validity, language validity, structural validity, and concurrent validity, as well as good internal consistency.

Chiu and Henry (1990): Mathematics Anxiety Scale for Children

This questionnaire is applicable for a wider age range than the MARS-E, as it has been successfully used with students aged between 9 and 14 years old. The questions on the Mathematics Anxiety Scale for Children (MASC) were also based on the MARS but shortened extensively so that it was easier for younger students to complete. The MASC breaks Maths Anxiety down into four factors: Maths evaluation, learning, problem-solving, and teacher anxiety.

> Chui and Henry (1990) put the MASC to the test and found that it has a good internal structure and a alpha value of 0.92, indicating that it is a reliable measure of Maths Anxiety in children.

The 22 questions in the MASC assess children's feelings towards mathematical situations but use a 4-point numerical Likert scale, rather than the most commonly used 5-point Likert scale. The choices are (1) not nervous, (2) a little bit nervous, (3) very nervous, or (4) very, very nervous. Having only four choices perhaps makes it easier for younger children to understand, as five choices is quite a lot, but four can still lead to some confusion and uncertainty, especially in the numerical form.

Gierl and Bisanz (1995): Mathematics Anxiety Survey

This questionnaire utilised a pictorial rating scale and included only 14 questions, making it easy to administer. The questions relate to attitudes towards Maths, Maths Anxiety, and anxiety in mathematical tests and offer five pictorial responses for the students to select from. It is suggested that the Mathematics Anxiety Survey (MAXS) can be used effectively with students aged between 8 and 13 years old.

> Gierl and Bisanz (1995) assessed the internal reliability of the MAXS. They reported a Cronbach's alpha of 0.85 for Grade 3 students (8–9 year olds) and 0.87 for Grade 6 students (aged 11–12 years old).

Jameson (2013): Children's Anxiety in Math Scale

This questionnaire was designed for use with children as young as six years old and asks only 16 questions. The questions relate to general Maths Anxiety, Maths Performance Anxiety, and Maths Error Anxiety. Using a pictorial response, children are asked to select the best facial image on the scale that represents their emotion to each situation posed to them. As this was developed in 2013, it is more up to date than a lot of available questionnaires, and perhaps a better representation of what children are learning in the classroom. It has been shown to be a reliable measure of children's Maths Anxiety through statistical testing.

> Primi et al. (2020) used EFA to investigate the dimensions measured within the Children's Anxiety in Math Scale. Three factors were found through this approach, namely General Maths Anxiety, Maths Performance Anxiety, and Maths Error Anxiety.

Thomas and Dowker (2000): Maths Anxiety Questionnaire

Thomas and Dowker aimed to design a self-reporting tool to measure Maths Anxiety in children aged between six and nine years old. The 24 questions in this self-reporting tool are categorised into four different areas: self-perceived performance, attitudes in Mathematics, unhappiness related to problems in Mathematics, and anxiety related to problems in Mathematics. Again, this questionnaire uses a pictorial Likert Scale to help younger children understand the 5-point scale.

> Krinzinger et al. (2007) adapted the MAQ for use in Germany and called it the FRA (Fragebogens für Rechenangst). Translated, this means "Questionnaire for Fear of Arithmetic." Cronbach's alpha for this questionnaire was between 0.83 and 0.91 and successfully helped them to measure Maths Anxiety, self-perceived performance and attitudes, as well as calculation ability. It was suggested that it was suited for use in clinical settings.

Wu et al. (2012): Scale for Early Maths Anxiety

The 20 questions within this questionnaire relate to anxiety experienced in both learning Maths, and within being in mathematical situations. The 5-point Likert scale was also pictorial and used emojis that illustrated varying degrees of worry, rather than sadness or happiness, which may make it easier to interpret than other pictorial scales.

> Wu et al. (2012) measured the reliability of the Scale for Early Maths Anxiety (SEMA). They found that the questionnaire had a Cronbach's alpha value of 0.87, demonstrating strong reliability.

The SEMA was developed specifically for children aged 7–9 years old, so the questions are reflective of what they may be learning in Maths lessons and what situations they may encounter. While the questionnaire is celebrated because it is specific in nature, it was developed in relation to the US curriculum, so it would need adapting for use in different countries and different education systems. For example, a question in the SEMA asks students how anxious they would feel when answering this question: How much money does Annie have if she has two dimes and four pennies? Children in other countries may feel confused with this question, so it may need to be adapted to ensure it is applicable to other cultures and countries.

> Kirkland (2020) used the SEMA to measure Maths Anxiety in Key Stage 2, with students aged between 7 and 11 years old. It was reflected that the SEMA, both in the pilot study and the main study, was easy to administer and accessible for all the participants within the research.

Petronzi et al. (2019): Children's Mathematics Anxiety Scale UK

Petronzi and colleagues developed this measurement tool in 2018, to use with children aged 4-7 years old, primarily in the UK. This age range is younger than other questionnaires available. The 19-item Children's Mathematics Anxiety Scale UK (CMAS-UK) also used a 3-point pictorial Likert scale to help young children understand the varying degree of the available responses. Having only three options to choose from for each scale may make the questionnaire more accessible, and understandable, for younger children, and the results more accurate.

> Petronzi et al. (2019) aimed to further validate the CMAS-UK, through testing its validity and reliability with 163 children, between the ages of 4 and 7 years old, across 2 primary schools in the UK. They reported high internal consistency with Chronbach's alpha of 0.88 and recommended its utility as a Maths Anxiety measurement tool for young children.

Qualitative Approaches

If you are looking to understand someone's experience of Maths Anxiety in a deeper way, there are different methods that you can use instead of, or alongside questionnaires.

There are qualitative ways to explore Maths Anxiety, which means collecting data that is based on words and meanings, rather than being numerical in nature. These methods can provide richer, realistic data but won't necessarily give you a "measure" of someone's Maths Anxiety. Nonetheless, by observing individuals or discussing their Maths Anxiety with them, you are more likely to have a deeper understanding of what they are experiencing, which might give greater insight into where, why, and when it may occur.

> Dove, Montague, and Hunt (2021) suggested that questionnaires may not fully capture the experience of Maths Anxiety. Instead, to examine primary school teachers' own experiences of Maths Anxiety, they conducted interviews and analysed them using Interpretative Phenomenological Analysis, which is a way in which to analyse the data collected and explore how individuals make sense of their experiences. They reported that doing so gave them specific detail, and a deeper understanding of the experience and recommended that future research should utilise this approach as well.

Observing individuals who are maths anxious may show you certain behaviours or behavioural patterns that you might not have been aware of before. As their teacher, you will be aware in general of what they do in your lesson, but looking at one student's reaction to Maths in particular during a whole lesson can be eye-opening. There may be micro-behaviours or patterns that form that you would not have been able to see while teaching the whole class, such as fidgeting or averting eye contact when the class are being asked questions, or biting their lip or frowning when they are finding something difficult.

> **Ben's Experience**
>
> During Maths lessons, Ben was usually quiet and didn't participate regularly. His teacher was aware of this and knew he was quite a passive student. During times when the teacher was asking questions to the class, he often rubbed his fingers together under the table or rubbed the edge of his chair, repetitively, with a frown on his face. His teacher hadn't noticed this before and was quite shocked as she hadn't thought him to be nervous. This shows how taking the time to watch someone within a situation, like a Maths lesson for example, can alert you to how they are feeling through observing small behaviours which might otherwise go unnoticed. Ben's teacher was keen to see whether his behaviours were repeated in other lessons or whether they were specific to his Maths lessons.

On top of this, such observations may give you a clearer picture of the situations or mathematical content that creates a rise in a student's anxiety. Reflecting in detail about what situations this happens in, such as when they are learning about division, or when they are answering questions in front of the class, will allow you to create ways to help them overcome their anxiety. Be wary though, because if the student is aware that they are being observed, their behaviour is unlikely to be natural and it may affect what you see.

> Kirkland (2020) used both quantitative and qualitative measures to explore Maths Anxiety. Using questionnaires to measure levels of Maths Anxiety before and after interventions took place provided an understanding of any change, while using observations and interviews together provided a more detailed understanding of what the students' experience was like, as well as useful information for the personalised interventions that occurred.

Similarly, talking to the individual will give you a deeper understanding of what they may experience, and perhaps how best to support them. This can be done in groups or on an individual basis, and you should think about whether you want this to be formal or informal. Formally interviewing someone is likely to increase their anxiety, and they may not be honest about their emotions. Their relationship with you will also affect their likelihood of talking openly, such as whether you have only just become their teacher, or whether they have known you for a few years. In a relaxed setting, reflective of an informal discussion, it is more likely that they will feel comfortable about sharing their thoughts and opinions, but this doesn't work for everyone. Sharing their genuine thoughts and emotions may prove tricky for some as well if they are in a group setting, so knowing the individual well beforehand, and what may be the most comfortable setting for them to help you understand their experience, is vital. Again, do what is best for you and the individual whose experience you want to explore.

> **Abdul's Experience**
>
> In one-on-one interviews, Abdul felt comfortable sharing his thoughts about Maths, including how it made him anxious, how he felt that he wasn't good enough to achieve his GCSE, and how he was always worried that his parents would be disappointed. When discussing Maths as part of a small group of four students, he didn't share his thoughts unless he was asked directly, and even then, his comments were less detailed and didn't represent the same level of anxiety compared to when he was one-on-one with the researcher. This illustrates the difference that interviews and group discussion can have on findings.

Interviews with other people who are prominent in the maths anxious individual's life can also be insightful, such as their parents, siblings, or teacher. This can provide further perspective on the individual's experience and help you gain a more holistic understanding, but it is important to note that everyone has their own perceptions, which are subjective, so relying on what others report may not always provide you with the most accurate picture.

So... Can I Measure It?

- Physiological and brain activity measures have been used to understand the nature of Maths Anxiety, but this requires expertise and equipment beyond the realms of the classroom. These also have limitations regarding the extent to which they are direct measures of Maths Anxiety.
- There are a range of questionnaires available which can indicate the extent of a person's Maths Anxiety, but they may differ in the factors that they measure.
- Questionnaires for children have also been developed but there are less available compared to those designed for adults and adolescents.
- Using qualitative approaches, such as observations or interviews, may lead to a deeper understanding of an individual's experience of Maths Anxiety.
- Don't be afraid to adapt measurement tools to suit your needs, but be aware that this may impact their reliability and validity.

If you are keen to measure Maths Anxiety, there are so many opportunities and ways for you to do so. While physiological and neurological measures are not achievable in the day to day classroom, knowing more about them may help you to spot behaviours related to Maths Anxiety, such as changes in breathing, eye movements, or perspiration. More suited to the classroom are questionnaires, which have been designed for adults and adolescents, as well as children. These vary in length, accessibility, the factors they measure, and their applicability to different contexts.

Alternatively, taking the time to observe a maths anxious individual may provide you with a deeper picture of their Maths Anxiety, as might talking with them, whether through formal interviews or informal discussions, as individuals or in groups. Doing this alongside questionnaires may give you a whole armoury of information in which to act on through a mixed methods approach; it doesn't have to be one or the other.

Whichever way you decide to learn more about the experience of Maths Anxiety, whether it be with an individual, a group, or a class, it is important to think about your next steps and know what is available for you to use. Your end goal is undoubtedly to help the person, or people, experiencing Maths Anxiety, so the way in which you collect data should help build you a platform from which to work. Remember, all existing tools to measure Maths Anxiety are adaptable and should be used to suit your context. Selecting the most appropriate one in the first place, however, is key.

References

Atkinson, R. T. (1988) Exploration of the Factors Relating to the System of Mathematics Anxiety. Masters Thesis. University of Oklahoma.

Baloglu, M. and Balgalmis, E. (2010) 'The Adaptation of the Mathematics Anxiety Rating Scale-Elementary Form into Turkish, Language Validity, and Preliminary Psychometric Investigation', *Educational Sciences: Theory and Practice*, 10(1), pp. 101–110.

Capraro, M. M., Capraro, R. M. and Henson, R. K. (2001) 'Measurement Error of Scores on the Mathematics Anxiety Rating Scale Across Studies', *Educational and Psychological Measurement*, 61(3), pp. 373–386. Available at: https://doi.org/10.1177/00131640121971266.

Chiu, L. H. and Henry, L. L. (1990) 'Development and Validation of the Mathematics Anxiety Scale for Children', *Measurement and Evaluation in Counseling and Development*, 23(3), pp. 121–127.

Cipora, K., Szczygieł, M., Willmes, K. and Nuerk, H. C. (2015) 'Math Anxiety Assessment with the Abbreviated Math Anxiety Scale: Applicability and Usefulness: Insights from the Polish Adaptation', *Frontiers in Psychology*, 6, pp. 1–16. Available at: https://doi.org/10.3389%2Ffpsyg.2015.01833.

Conroy, D. E. (2001) 'Progress in the Development of a Multidimensional Measure of Fear of Failure: The Performance Failure Appraisal Inventory (PFAI)', *Anxiety, Stress, & Coping*, 13, pp. 431–452. Available at: https://doi.org/10.1080/10615800108248365.

Demedts, F., Cornelis, J., Reynvoet, B., Sasanguie, D. and Depaepe, F. (2023) 'Measuring Math Anxiety Through Self-Reports and Physiological Data', *Journal of Numerical Cognition*, 9(3), pp. 380–397. Available at: https://doi.org/10.5964/jnc.9735.

Dove, J., Montague, J. and Hunt, T. E. (2021) 'An Exploration of Primary School Teachers' Maths Anxiety Using Interpretative Phenomenological Analysis', *International Online Journal of Primary Education (IOJPE)*, 10(1), pp. 32–49.

Dreger, R. M. and Aiken, L. R. (1957) 'The Identification of Number Anxiety in a College Population', *Journal of Educational Psychology*, 48, pp. 344–351. Available at: https://doi.org/10.1037/h0045894.

Elliot, A. J. and Murayama, K. (2008) 'On the Measurement of Achievement Goals: Critique, Illustration, and Application', *Journal of Educational Psychology*, 100, pp. 613–628. Available at: https://doi.org/10.1037/0022-0663.100.3.613.

Fennema, E. and Sherman, J. A. (1976) 'Fennema-Sherman Mathematics Attitudes Scales: Instruments Designed to Measure Attitudes Toward the Learning of Mathematics by Females and Males', *Journal for Research in Mathematics Education*, 7(5), pp. 324–326. Available at: https://doi.org/10.2307/748467.

Firouzian, F., Fadaei, M., Ismail, Z., Firouzian, S. and Yusof, Y. M. (2015) 'Relationship of Mathematics Anxiety and Mathematics Confidence Among Engineering Students', *Advanced Science Letters*, 21(7), pp. 2400–2403.

Freeston, M. H., Ladouceur, R., Letarte, H., Thibodeau, N. and Gagnon, F. (1991) 'Cognitive Intrusions in a Non-Clinical Population. I. Response Style, Subjective Experience, and Appraisal', *Behaviour Research and Therapy*, 29, pp. 585–597. Available at: https://doi.org/10.1016/0005-7967(91)90008-q.

Gierl, M. J. and Bisanz, J. (1995) 'Anxieties and Attitudes Related to Mathematics in Grades 3 and 6', *Journal of Experimental Education*, 63(2), pp. 139–158. Available at: https://doi.org/10.1080/00220973.1995.9943818.

Hopko, D. R., Mahadevan, R., Bare, R. L. and Hunt, M. K. (2003) 'The Abbreviated Math Anxiety Scale (AMAS) Construction, Validity, and Reliability', *Assessment*, 10(2), pp. 178–182. Available at: https://doi.org/10.1177/1073191103010002008.

Hunt, T. E., Bagdasar, O., Sheffield, D., and Schofield, M. (2019) 'Assessing Domain Specificity in the Measurement of Mathematics Calculation Anxiety', *Education Research International*, 2019, pp. 1–7. Available at: https://doi.org/10.1155/2019/7412193.

Hunt, T. E., Bhardwa, J. and Sheffield, D (2017) 'Mental Arithmetic Performance, Physiological Reactivity and Mathematics Anxiety Amongst UK Primary School Children', *Learning and Individual Differences*, 57, pp. 129–132. Available at: https://doi.org/10.1016/j.lindif.2017.03.016.

Hunt, T. E., Clark-Carter, D. and Sheffield, D. (2011) 'The Development and Part Validation of a UK Scale for Mathematics Anxiety', *Journal of Psychoeducational Assessment*, 29(5), pp. 455–466. Available at: https://doi.org/10.1177/0734282910392892.

Hunt, T. E. and Maloney, E. A. (2022) 'Appraisals of Previous Maths Experiences Play an Important Role in Maths Anxiety', *Annals of the New York Academy of Sciences*, 15(1), pp. 143–154. Available at: https://doi.org/10.1111/nyas.14805.

Hunt, T. E., Sheffield, D. and Clark-Carter, D. (2014) 'Exploring the Relationship between Mathematics Anxiety and Performance: An Eye-tracking Approach', *Applied Cognitive Psychology*, 29(2), pp. 226–231. Available at: https://doi.org/10.1002/acp.3099.

Jameson, M. M. (2013) 'The Development and Validation of the Children's Anxiety in Math Scale', *Journal of Psychoeducational Assessment*, 31(4), pp. 391–395. Available at: https://doi.org/10.1177/0734282912470131.

Kazelskis, R. and Reeves-Kazelskis, C. (1999) The Math Anxiety Questionnaire: A Simultaneous Confirmatory Factor Analysis across Gender. Alabama: Mid-South Educational Research Association.

Kirkland, H. R. (2020) An Exploration of Maths Anxiety and Interventions in the Primary Classroom. Ph. D. Thesis. University of Leicester.

Kooken, J., Welsh, M. E., McCoach, D. B., Johnston-Wilder, S. and Lee, C. (2016) 'Development and Validation of the Mathematical Resilience Scale', *Measurement and Evaluation in Counseling and Development*, 49, pp. 217–242. Available at: https://doi.org/10.1177/0748175615596782.

Krinzinger, H., Kaufmann, L., Dowker, A., Thomas, G., Graf, M., Nuerk, H. C. and Willmes, K. (2007) 'German Version of the Math Anxiety Questionnaire (FRA) for 6-to 9-Year-Old Children', *Zeitschrift fur Kinder-und Jugendpsychiatrie und Psychotherapie*, 35(5), pp. 341–351. Available at: https://doi.org/10.1024/1422-4917.35.5.341.

Lim, S. Y. and Chapman, E. (2013) 'Development of a Short Form of the Attitudes Toward Mathematics Inventory', *Educational Studies in Mathematics*, 82, pp. 145–164. Available at: https://doi.org/10.1007/s10649-012-9414-x.

Mattarella-Micke, A., Mateo, J., Kozak, M. N., Foster, K. and Beilock, S. L. (2011) 'Choke or Thrive? The Relation Between Salivary Cortisol and Math Performance Depends on Individual Differences in Working Memory and Math-Anxiety', *Emotion*, 11(4), p. 1000. Available at: https://doi.org/10.1037/a0023224.

OECD (2014) PISA 2012 Technical Report. Paris: OECD Publishing. Available at: https://www.oecd.org/pisa/pisaproducts/PISA-2012-technical-report-final.pdf (Accessed: 4 May 2024).

Pekrun, R., Goetz, T., Frenzel, A. C., Barchfeld, P. and Perry, R. P. (2011) 'Measuring Emotions in Students' Learning and Performance: The Achievement Emotions Questionnaire (AEQ)', *Contemporary Educational Psychology*, 36, pp. 36–48. Available at: https://doi.org/10.1016/j.cedpsych.2010.10.002.

Petronzi, D., Staples, P., Sheffield, D., Hunt, T. E. and Fitton-Wilde, S. (2019) 'Further Development of the Children's Mathematics Anxiety Scale UK (CMAS-UK) for Ages 4-7 Years', *Educational Studies in Mathematics*, 100(3), pp. 231–249.

Pletzer, B., Kronbichler, M., Nuerk, H. C. and Kerschbaum, H. H. (2015) 'Mathematics Anxiety Reduces Default Mode Network Deactivation in Response to Numerical Tasks', *Frontiers in Human Neuroscience*, 9, p. 202. Available at: https://doi.org/10.3389/fnhum.2015.00202.

Primi, C., Donati, M. A., Izzo, V. A., Guardabassi, V., O'Connor, P. A., Tomasetto, C. and Morsanyi, K. (2020) 'The Early Elementary School Abbreviated Math Anxiety Scale (the EES-AMAS): A New Adapted Version of the AMAS to Measure Math Anxiety in Young Children', *Frontiers in Psychology*, 11, p. 1014. Available at: https://doi.org/10.3389/fpsyg.2020.01014.

Richardson, F. C. and Suinn, R.M. (1972) 'The Mathematics Anxiety Rating Scale: Psychometric Data', *Journal of Counseling Psychology*, 19(6), pp. 551-554. Available at: https://doi.org/10.1037/h0033456.

Sachs, J. and Leung, S.O. (2007) 'Shortened Versions of Fennema-Sherman Mathematics Attitude Scales Employing Trace Information', *Psychologia*, 50(3), pp. 224-235.

Suinn, R. M., Taylor, S. and Edwards, R. W. (1988) 'Suinn Mathematics Anxiety Rating Scale for Elementary School Students (MARS-E): Psychometric and Normative Data', *Educational and Psychological Measurement*, 48, pp. 979-986. Available at: https://doi.org/10.1177/0013164488484013.

Suinn, R. M. and Winston, E. H. (2003) 'The Mathematics Anxiety Rating Scale, a Brief Version: Psychometric Data', *Psychological Reports*, 92(1), pp. 167-173. Available at: https://doi.org/10.2466/pr0.2003.92.1.167.

Taylor, J. A. (1953) 'Taylor Manifest Anxiety Scale (TMAS)', *Journal of Abnormal Psychology*, 48(2), pp. 285-290. Available at: https://doi.org/10.1037/t00936-000.

Thomas, G. and Dowker, A. (2000) Mathematics Anxiety and Related Factors in Young Children. Bristol: British Psychological Society.

Vallerand., R. J., Pelletier, L. G., Blais, M. R., Brière, N. M., Senécal, C. B. and Vallières, E. F. (1992) 'The Academic Motivation Scale: A Motivation Scale: A Measure of Intrinsic, Extrinsic, and Motivation in Education', *Educational and Psychological Measurement*, 52, pp. 1003-1017. Available at: https://dx.doi.org/10.1177/0013164492052004025.

Wigfield, A. and Meece, J. L. (1988) 'Math Anxiety in Elementary and Secondary School Students', *Journal of Educational Psychology*, 80(2), pp. 210-216. Available at: https://doi.org/10.1037/0022-0663.80.2.210.

Wu, S. S., Barth, M., Amin, H., Malcarne, V. and Menon, V. (2012) 'Math Anxiety in Second and Third Graders and Its Relation to Mathematics Achievement', *Frontiers in Psychology*, 3, p. 162. Available at: https://doi.org/10.3389/fpsyg.2012.00162.

Young, C. B., Wu, S. S. and Menon, V. (2012) 'The Neurodevelopmental Basis of Math Anxiety', *Psychological Science*, 23, pp. 492-501. Available at: https://doi.org/10.1177%2F0956797611429134.

4 What's Involved?

Researchers have identified commonalities across people's experiences of Maths Anxiety, shedding light on the factors that are likely at play. However, everyone's experiences in life are unique to them, and the way in which people encounter Maths Anxiety is no exception. No two experiences of Maths Anxiety will be exactly the same, as we have all been shaped differently and respond to various mathematical situations in our own way.

This chapter will explore the factors that researchers have found to be present within the experience of Maths Anxiety, when someone who is maths anxious is using or learning Maths. While this list is not definitive, the following factors have been recognised as being part of many experiences:

ATTAINMENT: measured results, typically considered to reflect a person's ability to understand and learn mathematical concepts, and to put this into practice effectively through testing.

WORKING MEMORY: part of the cognitive system responsible for temporarily holding and manipulating information needed for tasks, such as solving problems or making decisions.

ATTITUDES: the feelings and beliefs that someone has in relation to other people, situations, and objects, which can influence their behaviour and decisions.

SELF-BELIEF: a person's confidence and belief in their own abilities, potential, and worth.

MOTIVATION: someone's inner drive that makes them to take action, persevere, and pursue their goals.

AVOIDANCE: behaviour that allows someone to evade a situation, or a person, in order to prevent discomfort and negative consequences.

> **Izzy's Experience**
>
> Izzy's Maths Anxiety was related to low attainment, as she found Maths challenging. She regularly engaged in negative self-talk about her abilities, showing that she had a lack of self-belief. Izzy identified the importance of needing to learn Maths and didn't show a lack of motivation, but she did try to avoid Maths when possible, such as finishing her work quickly and not completing her homework unless prompted.

Someone who is maths anxious may relate to all the factors discussed in this chapter, most of them, or only a few; it is personal to them and their experience. Understanding the commonalities related to Maths Anxiety, however, not only aids in identifying it in educational settings but also enhances our understanding of the individual's perspective and what they are experiencing. This will hopefully lead to a more tailored response in helping maths anxious individuals to develop a more positive reaction to learning and using Maths, grounded in the acknowledgement that Maths Anxiety has a unique, personal nature, rather than applying a "one-size-fits-all" approach.

> Stoehr (2017) conducted longitudinal research over an 18-month period as part of a multi-layer study called "TEACH Math" in the US. Focusing on three university students in detail, it was found that each had their own distinct fears and coping mechanisms in relation to Maths Anxiety. Although there were similarities within the experiences, the resulting article was entitled "One Size Does Not Fit All," highlighting the unique nature of Maths Anxiety.

Attainment

When educators refer to attainment, they are not typically discussing the progress a student has made over a period of time, which is often classed as "achievement." Instead, the phrase "attainment," sometimes used interchangeably with the term "performance," is commonly used to describe learning outcomes, most likely in the form of objective, numerical results achieved at the end of a period of learning. This is often how cumulative success and accomplishments are measured in education, which can happen at various points throughout a student's academic journey. Whether on a regular or infrequent basis, standardised assessments, often referred to as "exams" or "tests," are typically used to indicate whether a student is working below, at, or above age-related expectations, using predefined criteria.

In Maths, measuring a student's attainment can be really valuable. This assessment can provide useful insights into the specific mathematical skills and concepts that individual students have mastered, or are finding difficult, enabling teachers to tailor the student's level of support, or challenge, accordingly. In addition to its usefulness on an individual level, measuring mathematical attainment can contribute to a broader understanding of the successes of teaching and learning Maths within a class, across the school, or on a national level. In the UK, for example, mathematical attainment is compared nationally through the standardised testing of the Key Stage 2 SATs (Standard Assessment Tests). These standardised tests provide comprehensive data about individual students' mathematical attainment compared to the national average, as well as specific demographic groups and the school as a whole. Mathematical attainment is also compared globally, through organisations like the Organisation for Economic Co-operation and Development's Programme for International Student Assessment as we discussed earlier.

The focus of attainment as a measure of mathematical success, however, does have its drawbacks. It is often argued that using standardised assessments to determine a student's mathematical attainment, and therefore ability, can lead to inequality, as not all students can access these methods of assessment. The majority of standardised assessments may also fail to provide all students with the opportunity to showcase their critical thinking or problem-solving skills, which are essential components of successfully using and learning Maths. Therefore, the use of such assessments to measure mathematical attainment can be restrictive, limiting and not reflective of a student's true mathematical ability.

On top of this, solely relying on objective measures of attainment in Maths can lead to a narrow and perhaps limited definition of what mathematical success actually is. Assessing a student's mathematical ability based on a numerical outcome does not illustrate the learning process, or the effort a student invested into learning and understanding particular concepts that they found challenging. Many things can also adversely affect this snapshot understanding of a student's attainment, such as a lack of sleep, or struggling with low motivation, leading to poorer outcomes than expected. This may suggest that standardised assessments of mathematical attainment are not always accurate, as they can be negatively impacted by outside factors.

> Aunola, Leskinen, and Nurmi (2006) used a cross-lagged longitudinal research design and found that the mathematical attainment of 196 children in Finnish primary schools was interlinked with their learning motivation and interest in Maths. If they felt positive towards Maths and were motivated, they were more likely to have high attainment compared to their peers.

But how does this relate to Maths Anxiety?

Studies have shown that it is more likely for a maths anxious individual to have low attainment compared to those who are not maths anxious. For a maths anxious student, they may be likely to find learning and using Maths challenging, and not be able to demonstrate an understanding of what is being taught or learnt in lessons. They may consistently perform poorly on tests, irrespective of whether they find timed situations anxiety-evoking, and find it difficult to comprehend new mathematical concepts, or recall previously taught concepts.

Amy's Experience

Throughout primary school, Amy's mathematical attainment, as tracked and measured by end-of-year assessments, decreased. She began Year 2, when she was six years old, with age-related expectations. This declined across the years to her being below age-related expectations at the end of Year 6 SATs. Amy shared that she often worried that she found Maths really hard, and that she spent most of the lesson worrying rather than focusing on what she needed to learn – which made her worry more!

It has also been suggested that there is a potential relationship between Maths Anxiety and mathematical difficulties, such as Dyscalculia, a specific learning difficulty that affects an individual's ability to comprehend, process, and use mathematical concepts successfully. Both Maths Anxiety and Dyscalculia can negatively impact a student's ability to learn and use Maths and can overlap in the impact that they have. However, both have been found to be separate and independent constructs, despite being similar in their effect. Dyscalculia is often seen as a learning difficulty that negatively impacts learning and using Maths, while Maths Anxiety is seen as a negative emotional response to learning and using Maths. It is possible for a maths anxious student to also experience Dyscalculia, and vice versa. Maths Anxiety is generally higher in people with Dyscalculia, but this does not mean Maths Anxiety is necessarily higher in every person with Dyscalculia.

As low attainment is regularly associated with Maths Anxiety, researchers have aimed to identify whether Maths Anxiety can result from an individual having low mathematical attainment, or whether Maths Anxiety can lead to low mathematical attainment. Knowing the direction of this relationship could provide a deeper understanding of the experience of Maths Anxiety, and perhaps why it may develop, or what it may lead to. However, this is a heavily-debated topic: does low attainment lead to Maths Anxiety, or does Maths Anxiety lead to low attainment? Is there any way to actually know?

Extensive research investigating the relationship between Maths Anxiety and attainment is correlational in nature. Correlational research investigates relationships and can identify if different factors are connected to each other, such as Maths Anxiety and attainment. This type of research cannot identify why these connections exist though, it can only identify trends and patterns. Due to this, we cannot be certain whether Maths Anxiety impacts mathematical attainment, or low mathematical attainment can lead to Maths Anxiety, or if both of these claims are true.

Some correlational research into the relationship between Maths Anxiety and attainment is longitudinal. This means that the research, although still unable to determine cause and effect, is conducted over a longer period of time, leading to a deeper understanding of how the relationship may develop.

For example, attainment and Maths Anxiety may be measured when participants are ten years old. Over the next three years, both constructs may be measured every six months, until the participants are 13. This then gives a trajectory of both Maths Anxiety and attainment. Imagine a student has low mathematical attainment at ten years old, and low levels of Maths Anxiety. Their attainment remains low across the three years while their Maths Anxiety steadily increases. This may indicate that their low attainment preceded, and led to, Maths Anxiety.

While this may be the case, many other extraneous factors could have led to the development of Maths Anxiety over the three years, rather than low attainment, such as negative experiences in Maths or declining levels of motivation, for example. As Maths Anxiety is complex and encompasses a multitude of factors, it is difficult to isolate any one factor on their own, which is why it is so difficult to confirm causality.

One argument about the directionality between Maths Anxiety and attainment is based upon the Deficit Theory, which suggests that Maths Anxiety develops because of poor mathematical knowledge and understanding, possibly even a deficit in basic numerical processing. For example, if a student is repeatedly scoring 3/10 in their Maths tests, even though they are trying really hard to improve their attainment, this can negatively impact their relationship with Maths and lead to the development of Maths Anxiety. There has been a lot of support for this theory from research exploring Maths Anxiety experienced by young children, all the way to university students.

> Maloney, Ansari, and Fugelsang (2011) suggest that Maths Anxiety might result from a basic deficit in numerical processing, which could compromise the development of higher level mathematical skills.

Alternatively, the Debilitating Anxiety Model suggests that Maths Anxiety has a negative effect on a student's ability to learn mathematical concepts and to recall prior knowledge, therefore negatively impacting their ability to successfully meet mathematical goals set through the curriculum.

An example of this may be that during a Maths lesson, a student may be too preoccupied with feelings of anxiety and worry, which stops them from fully engaging with the lesson and learning, leading to poorer outcomes than if they were not focusing on their feelings of anxiety. While this may be interlinked with someone's working memory ability, which will be discussed later in this chapter, this indicates that Maths Anxiety can have a negative impact on mathematical attainment.

> Carey et al. (2016) note that evidence for the Debilitating Anxiety Model suggests Maths Anxiety can impact performance at the stages of pre-processing, processing, and retrieval of Maths knowledge. They also identified that neither the Deficit Theory nor the Debilitating Anxiety Model can fully explain the relationship between Maths Anxiety and Maths attainment, leading to the suggestion that more longitudinal, and mixed-methods, research should be conducted to explore the existence of what is likely to be a reciprocal relationship.

A further suggestion to consider is the Reciprocal Theory, which takes into account how both Maths Anxiety and low attainment may impact each other in a cyclical pattern.

> Pekrun (2006) explained the relationship between Maths Anxiety and mathematical attainment as a cognitive feedback loop: Maths Anxiety negatively affects mathematical attainment, which then increases the level of Maths Anxiety experienced. This pattern continues, with debilitating effects.

A maths anxious student may be unable to pay attention to the content being taught and find it difficult to learn in the Maths classroom as a result. This can lead to poor attainment as

Maths Anxiety has negatively impacted their ability to learn and use Maths. In turn, scoring low on a test and being aware of having low mathematical attainment can create the negative feelings and emotions related to Maths Anxiety, such as feeling inadequate, having low self-belief, and feeling incapable. This could be related to feeling pressured to have good academic outcomes or feeling that they are unlikely to successfully achieve in Maths assessments. This detrimental loop can impact both attainment and Maths Anxiety, leading to poorer academic outcomes and a negative relationship with Maths.

> **Sarah's Experience**
>
> Sarah had above-average attainment in Maths. Her parents shared that this was always the case, and that she consistently did well in assessments, and also in lessons. She did have a good grasp of mathematical concepts, but she was also highly maths anxious. Interestingly, Sarah felt as though she was "bad at Maths," but objectively speaking, she wasn't. This shows that not every maths anxious individual has low mathematical attainment, but it also highlights the influence of self-belief upon Maths Anxiety. Does someone's lack of self-belief have a greater connection to their feelings of anxiety compared to actual attainment?

While it can be the case that maths anxious students have high mathematical attainment, this appears to be a deviation away from the trend. It is important though, when assessing the link between Maths Anxiety and attainment, to look beyond a simple linear relationship. Researchers often investigate the existence of a general pattern between Maths Anxiety and attainment, which is typically done in two ways. Firstly, they might compare attainment between groups of people with particularly low and high Maths Anxiety. Secondly, they might assess the statistical correlation between the variables, based on maintaining the variables as continuous. Both of these approaches run the risk of missing out on more nuanced cases, such as students who buck the trend by having profiles that are valid but quite different to the majority. Such cases can sometimes be lost within datasets when the primary objective is to test for a general group difference or a general relationship.

A more complex, but often more informative, approach is to test for what are called "latent profiles," which helps researchers identify those students with different characteristics to the norm, for example, students with high attainment and high levels of Maths Anxiety.

> Cipora et al. (2023) surveyed 4,597 students at ETH Zurich, one of the most elite technical universities in the world. Even in this group, there was evidence of Maths Anxiety, albeit lower than in students in other studies. Small correlations with attitudes towards Maths suggested that being anxious about Maths does not preclude positive feelings towards it.

Therefore, when thinking about Maths Anxiety, it is extremely important not to overlook students who do not have low mathematical attainment, as this is not guaranteed to be an indication of Maths Anxiety. It's also important not to assume that a student with low attainment is maths anxious, yet acknowledging that low attainment can often be a factor within the experience of Maths Anxiety provides teachers with a valuable insight that can be used to identify and support maths anxious individuals in the classroom. We should also be mindful of students in the middle range of Maths attainment and anxiety, who are not in obvious need of support but run the risk of experiencing a downward trajectory. The situation is even more complex when we consider how student profiles can vary according to different forms of anxiety, including the way such profiles differentially relate to Maths performance.

> Rossi, Xenidou-Dervou, and Cipora (2023) studied anxiety profiles in 328 university students in the UK. In addition to the typical profiles (high or low anxiety across all types: Maths Anxiety, Test Anxiety, and General Anxiety), three more specific profiles were observed, which include different combinations of high, medium, or low anxiety across the different forms of anxiety. Students' Maths performance varied according to the profile group they belonged to.

It's also important to point out here that studies of changes in Maths Anxiety and Maths attainment are often based on absolute scores. That is, the final grade that might be achieved on a standardised test. They often neglect to take into account the bits in between tests, such as the additional time and effort that students put in, and the greater support offered by teachers. In other words, the reality of Maths education can be lost in such analyses, instead focusing on a simple, linear relation between Maths Anxiety and Maths attainment. It is possible that any observed relations are then simplified and potentially misrepresent reality, without taking into account the variation in students' experience as a result of baseline differences or changes over time. For instance, we know that Maths Anxiety is related to a range of other individual differences that can impact Maths learning and attainment, including Maths attitudes and self-beliefs. Taking into account so many relevant factors, with large samples over time, is usually not feasible. Nevertheless, studies involving big data are valuable and should not be overlooked, often providing evidence for the relative importance of variables from a developmental perspective.

Working Memory

Working memory is important in relation to both Maths attainment and Maths Anxiety. Working memory is like your mental post-it note, holding onto information that is crucial for you to complete tasks or solve problems. It is an executive function of the brain which helps us to process, use, and remember information over a short period of time, usually around two minutes. Executive functions (sometimes referred to as executive control or cognitive control) are "top-down" mental processes that are essential for all types of cognitive performance. They enable us to monitor our actions and allow us to solve problems and shift strategies flexibly, to ignore distractors, and inhibit irrelevant impulses. In the case of temporarily maintaining information in working memory, after a short period of time, it's processed and stored, or forgotten.

Imagine it's like the sorting station of a post office. All the letters, like the stimulus, arrive at the sorting station and are organised according to their need for attention, whether it's important or just junk mail. If a letter is deemed unimportant, just like the stimuli, it is promptly discarded. If the letter does need attention, it is sorted further to ensure it arrives at the office of the person who has the skills to efficiently attend to it.

> **The Working Memory Model**
>
> The Working Memory Model (Baddeley and Hitch, 1974) describes how stimuli are processed by our working memory. When incoming stimuli arrive in the brain, they are attended to by the central executive first, which is responsible for processing the stimuli, and directs attention where it's needed, makes decisions, prioritises where attention goes, retrieves memories related to the stimuli, and more. If stimuli need to be processed further, they are allocated to specialist subsystems: the phonological loop, the visuospatial sketchpad, and the episodic buffer. If the stimuli are spoken or written in nature, they will be processed by the phonological loop, while the visuospatial sketchpad deals with visual and spatial information. The episodic buffer is there to temporarily store and integrate information from the phonological loop and visuospatial sketchpad, as well as long-term memory, binding it together.

Working memory is a remarkable cognitive tool, but it does have limitations. It can only hold information for a brief duration of time, and if there is a large amount of incoming stimuli, it can become overloaded. There is a limit to what stimuli the working memory can hold, as well as process, which is why it is sometimes difficult to do numerous things at the same time, especially when the tasks tap into the same components of working memory, for example, if they are both phonological in nature. This is why it can be tricky to hold a phone conversation while someone else in the room is trying to talk to you at the same time. The capability and restrictions of working memory have prompted researchers to explore the relationship between working memory and attainment, namely mathematical performance, and also investigate whether Maths Anxiety can overwhelm and hinder the efficiency of the working memory.

> Friso-Van den Bos et al. (2013) found that all components of the working memory system are required to perform successfully on mathematical tasks.

Working memory plays a crucial role in Maths, particularly in relation to mathematical reasoning and problem-solving. Depending on the mathematical situation you are engaged in, it is likely that you will need to process and manipulate multiple pieces of information simultaneously, including numbers, symbols, and operations. Working memory holds these elements in your mind while you process and manipulate them to address and solve the mathematical question at hand.

For example, let's look at calculating 27 multiplied by 12. In this instance, your working memory would work to retain both numbers and the multiplication operation while you navigate the required steps of the calculation. Depending on the strategies you would use, this might include tasks like partitioning one of the numbers, such as 12 into 10 and 2, and calculating 27 multiplied by 10 to begin with. Your working memory would then also hold the interim answer of 270 in your mind while simultaneously solving 27 multiplied by 2, before adding both interim answers of 270 and 54 together to arrive at your answer of 324. Throughout this process, your working memory helps you remain on task and align your answer with the question given to you initially, which is being stored while you work on the task at hand.

Working memory also supports the acquisition of new mathematical knowledge and concepts, as it can help to interpret the relationships between concepts, procedures, and symbols, and connect new, incoming stimuli with related knowledge that is already embedded in your long-term memory. This facilitates your understanding of new mathematical concepts, as it relates incoming stimuli with the mathematical knowledge you already have, allowing you to apply your existing mathematical knowledge to the present learning situation, learn new concepts, and improve your mathematical proficiency.

> Faust, Ashcraft, and Fleck (1996) explained that working memory can't support the demands of complex addition Maths questions if an individual is anxious. They found that highly maths anxious students were slower and less accurate when solving addition questions that required them to carry or problem solve, compared to their less anxious peers.

Maths can often place high demands on working memory, especially when solving multi-step problems or learning new concepts. This can lead to an overwhelming workload, making it challenging for working memory to process all of the incoming information simultaneously. When learning and using Maths, individuals do not only need process the mathematical content and the task at hand but also contend with external stimuli from their environment. These are often unrelated to Maths, such as background noise, conversations happening near them, and the movements of others around them. This may mean someone struggles to stay focused on the task at hand, as their working memory is attending to too much information all at once, and cannot direct its attention where required. As a result, individuals may find it hard to stay focused on the task at hand and achieve what is required of them.

The amount of information that working memory can hold and process at once, its capacity, can differ for each individual. This can develop over time. However, some people have a low working memory capacity, while some may have a high working memory capacity, regardless of age. This means that our ability to retain and process incoming stimuli can differ from those around us.

The capacity of our working memory can impact our ability to solve mathematical tasks successfully. It has been suggested that individuals with a high working memory capacity are better equipped to handle a greater volume of incoming stimuli and process it effectively. If someone has a high working memory capacity, they are likely to use their working memory regularly as it can lead to success when solving mathematical problems. This repeated use can, in turn, contribute to a positive cycle, strengthening the efficiency of their working memory even further. However, it is likely those with a higher working memory capacity have no other strategies to utilise when their working memory does become overwhelmed, as they are often reliant on their working memory working successfully and helping them achieve in Maths. This is one of the arguments put forward by some researchers who have observed a greater drop in Maths performance when someone has both high Maths Anxiety and high working memory capacity.

On the other hand, if someone has a low working memory capacity, they may encounter challenges in retaining and processing information. This may mean that they are less likely to utilise their working memory and, instead, rely on other methods to offload the cognitive burden, such as writing down information on a piece of paper. This allows this information to "leave" their working memory, freeing it up to process and retain other information that might be needed.

Jon's Experience

Although Jon's working memory wasn't assessed, he often made notes throughout lessons on his mini whiteboard. These were related to what the teacher was explaining or demonstrating, but also the calculations he was solving independently. When asked why, he said that it helped him think.

Using alternative strategies to overcome the limitations of a low working memory capacity has been found to be effective, allowing those individuals to successfully solve mathematical problems. If someone with a low working memory capacity is unable to use other strategies though, they may find it challenging to process and retain the information needed to learn and use Maths successfully, as their working memory can become quickly overwhelmed. This can negatively impact their mathematical attainment.

Ashcraft and Kirk (2001) measured Maths Anxiety and working memory capacity in 66 university students and asked them to mentally calculate the answers to one- and two-column addition questions. Half of the questions required numbers to be carried, as part of a dual-task procedure (the second task being to recall random letters at the same time). They found that the addition problems that involved carrying led to lower accuracy in highly maths anxious individuals, compared to those who were less anxious, especially as the second task became more difficult. They suggested this was due to anxiety disrupting central executive processes in particular.

Regardless of whether someone has a high or low working memory capacity, their working memory can become overwhelmed and, if they don't have an alternative strategy to support them in solving the mathematical task at hand, their attainment can be negatively affected. In particular, mathematical tasks involving mental arithmetic or those that require more than one step to arrive at the answer can be the most challenging, as they demand a lot of working memory resources. Therefore, it can be useful to identify those individuals who are struggling with their working memory, particularly in the Maths classroom, to support their learning and provide alternative strategies to help them retain and process incoming stimuli.

If someone's working memory is overwhelmed and not working effectively, they may find it difficult to follow instructions given to them in a single chunk. Instead, they may find it more manageable to receive instructions given in a step-by-step fashion, as this places less pressure on their working memory. An individual may also stop working on the task at hand because they lose track of what is required of them, and they may often appear to be inattentive or "daydreaming." They can also seem to be easily distracted by external stimuli around them, such as their classroom or classmates, rather than actively engaging in the learning process.

Among researchers, it is widely recognised that working memory and Maths Anxiety are linked. A common explanation for the negative impact of Maths Anxiety on Maths performance, particularly arithmetic performance, is that Maths Anxiety depletes working memory resources – the very resources we need to successfully attend to, process, and successfully complete mathematical problem-solving.

> Eysenck and Calvo's (1992) Processing Efficiency Theory (usually referred to as PET) has been used for several years by academics wanting to explain why Maths Anxiety often predicts poorer Maths performance, especially longer response times on Maths problems that require greater working memory resources. PET is a useful theoretical framework for explaining the interplay between anxiety and performance on cognitive tasks. This is why it is so popular in the context of Maths Anxiety research.

One argument is that, if someone is maths anxious, their anxiety uses up available working memory resources and directs it away from the task at hand and allocates attention to the anxiety occurring instead. If someone is tackling a multi-step word problem, for example, this requires their working memory to retain and process a range of information. If they are experiencing Maths Anxiety while doing this, their working memory will be allocating its attention to the individual's feelings of anxiety, leaving fewer available resources to successfully complete the multi-step word problem. This may lead to lesser academic outcomes, because they are not able to fully focus on what the question is asking of them. The individual may make mistakes in their calculations, arrive at the incorrect answer, lose focus and not recall what they were doing, or feel that they cannot answer it and therefore, do not attempt to. This suggests that Maths Anxiety can interact with working memory, potentially hindering someone's academic outcomes in Maths. It is through this process that Maths Anxiety may result in a sudden or unexpected deterioration in Maths performance.

> Ashcraft and Moore (2009) used the term "affective drop in performance" to describe how Maths Anxiety may lead to a decline in Maths performance, particularly in high-stake situations.

Some argue that the experience of Maths Anxiety impacts working memory capacity, as an individual is unable to retain and process the required stimuli, exacerbating their anxiety in mathematical situations. This can create a detrimental cycle of learning, where an inefficient working memory can trigger feelings of anxiety because of an individual's difficulty to attend to all the information required to solve a mathematical problem. This anxiety, in turn, depletes the available resources from the working memory, further reducing its effectiveness and impairing its ability to facilitate problem-solving, leading to intensified anxiety levels, in a perpetual loop. This highlights the interlinked nature of Maths Anxiety and working memory, and how this can have a negative impact on learning and using Maths.

> National Mathematics Advisory Panel (NMAP; 2008) suggested that a low working memory capacity was a potential risk factor for Maths Anxiety.

On the other hand, it is argued that individuals with a lower working memory capacity are actually less anxious in Maths because they are able to use a broader range of problem-solving strategies. In situations where their working memory becomes impaired by anxiety and unable to process the incoming stimuli, they can rely on alternative methods to answer the question and alleviate their anxiety, like a coping mechanism. Some researchers argue that this is in fact the reason why individuals with a higher working memory capacity are more likely to be maths anxious, as they over-rely on their working memory and feel anxious when it becomes overloaded and are left without alternative strategies. The range of interpretations suggests that, regardless of someone's working memory capacity, Maths Anxiety can still have a negative impact and overwhelm working memory, leading to adverse effects.

> Ramirez et al. (2013) suggested that individuals with a high working memory capacity and high level of Maths Anxiety may be more likely to have lower mathematical attainment because they rely heavily on their working memory to solve Maths tasks. When Maths Anxiety diminishes their attentional resources, this impacts their mathematical performance as they are used to relying on their working memory and do not have other resources available to them.

While there are conflicting theories regarding the specific working memory capacity and functioning capability held by maths anxious individuals, there is agreement that Maths Anxiety and working memory are related to one another. This is a valuable insight for teachers, as it helps us to identify how Maths Anxiety may hinder an individual's ability to tackle mathematical tasks that demand a high level of functioning from their working memory and reflect on the nature of tasks that are given.

> Vukovic et al. (2013) conducted longitudinal research into working memory and its relation to Maths Anxiety in 113 eight- and nine-year-old children. They found that higher levels of Maths Anxiety in the second grade predicted less improvement in mathematical attainment between the second and third grade, but only for students with high working memory capacity. This was not found for students with high Maths anxiety and lower working memory capacity.

While we have touched on the idea of working memory quite broadly, it is worth focusing on the concept of attentional processes here as well. Previously, Maths Anxiety researchers tended to only focus on the concepts of working memory capacity and the inefficiency of working memory when anxiety comes into play. More recently, there has been a shift towards the role of attention.

> Eysenck et al. (2007) developed Attentional Control Theory, which is an extension of the earlier Processing Efficiency Theory. While both these theories are general theories about the way in which anxiety might impact cognitive performance, they have both been used to explain Maths Anxiety effects. Attentional Control Theory focuses in particular on what we pay attention to when presented with a cognitive task, such as mathematical problem-solving.

Some Maths Anxiety researchers have explained their findings using this theory, based on the principle of goal-directed or stimulus-driven attentional systems. This seems to fit quite nicely with what researchers have been saying about Maths Anxiety for many years – that it is often associated with intrusive or worrisome thoughts. From a goal-directed perspective, some people seem to be quite good at focusing on the task at hand; they can inhibit their attention towards distracting information, whether that is in the environment or even their own thoughts. We may see this quite regularly in the classroom, where some students have high focus and can ignore the distractions of their classmates around them and focus on the task at hand.

On the other hand, anxiety can often result in someone directing their attention towards the things they are anxious about, especially when they first appear to them (what is sometimes referred to as "hypervigilance"). Even when they know distractions are not helpful, they might struggle to inhibit their attention to them. In the case of Maths Anxiety, this stimulus-driven attentional system might override the goal-directed system, meaning that someone attends to the negative, intrusive thoughts that they have. That is, someone may pay attention to hindering thoughts such as *"I can't do this"* or *"I'm going to make a mistake,"* which reduces their attention given to the Maths task they need to complete.

> Hunt, Clark-Carter, and Sheffield (2014) asked over 100 students to report what thoughts they had during a basic Maths task involving a series of mathematical verification problems. They found that students had lots of thoughts! Many of these thoughts were not related to completion of the task itself but rather focused on worries. For example, these included thoughts regarding panic, time pressure, making mistakes, and physical changes. Those who had particular negative thoughts were also significantly more maths anxious.

Similarly, external features of the environment that may be associated with pressure, anxiety, and fear could absorb someone's attention, again meaning that their attention is taken away from the Maths task at hand. Even seeing some algebraic equations on a board, or listening to someone talk about a Maths test could take someone's attention away from the Maths learning that they are supposed to be, and maybe even want to be, engaged in. This is quite important to acknowledge as teachers, as identifying ways that we can reduce external stimuli that might divert students' attention, whether this be visual or audio, may help to reduce the negative effect of Maths Anxiety.

> **Jodie's Experience**
>
> Jodie found it quite difficult to concentrate in Maths lessons. Initially, she said this was because she wasn't good at Maths, but upon further discussion, she believed that became easily distracted. She felt that everything around her was too chaotic, with lots of students talking about things mostly unrelated to Maths. This led her to become frustrated, which then led to thoughts about how she was finding it difficult, to the point of questioning what the point in trying was. All in all, Maths was starting to become an anxious experience for Jodie.

Attitudes

Attitudes are made up of our basic likes and dislikes, and our perceptions of people, situations, and things. Consciously or subconsciously, these attitudes guide our decisions and actions in everyday life, such as someone liking football for example, and therefore watching games regularly because they perceive it to be enjoyable. In relation to Maths, the term "attitudes" has been used quite variably. Some researchers use the term quite broadly, taking it to include emotions, often incorporating Maths Anxiety into this. However, Maths attitudes are generally accepted to be a person's opinion of Maths, including the level of enjoyment they associate with Maths. This also includes a person's general view of Maths, which usually involves their thoughts on its value or utility. Some researchers have measured quite specific attitudes towards Maths, such as the perception of Maths as a male domain and perceptions of the gendered nature of it. Sometimes, Maths attitudes measures incorporate self-belief (confidence, self-concept, self-efficacy, etc.) and motivation too, but it can be argued that these go beyond actual attitudes.

Everyone has an attitude towards Maths. Some people may have a positive attitude, where they may hold the opinion that Maths is interesting, useful, or easy to master. Alternatively, and much more researched, individuals may have a negative attitude towards learning and using Maths, with the belief that Maths is difficult, useless, or boring.

Attitudes towards Maths can affect people's behaviour and engagement. If someone has a positive attitude towards Maths, they are much more likely to engage with the subject, be motivated to succeed, and consistently put a high amount of effort into learning and using Maths, potentially leading to good outcomes. On the other hand, if an individual has a negative attitude towards Maths, this will likely prompt them to avoid it, and choose not to engage with the subject. In the classroom, these students are likely to be disengaged, avoidant when discussing and completing tasks, and will likely choose not to continue learning Maths when this option becomes available to them, which can negatively affect their future educational and career choices.

Some students might feel ambivalent towards Maths, experiencing mixed feelings. Mixed or contradictory feelings and thoughts about Maths might arise not only from a range of positive and negative experiences but also from different motivations and external pressures. For example, take a student who has always quite enjoyed Maths at school; they experience a sense of satisfaction from successful problem-solving and like the fact that they tend to do well in tests. However, their new Maths teacher seems to get cross and frustrated if they don't complete Maths problems at the rate the teacher expects, and the student's parents keep talking about progressing to Maths at a higher level; something that the student feels creates unnecessary pressure for them to do well. These combined experiences have the potential to generate conflicting attitudes towards Maths.

Alisha's Experience

Working to understand Alisha's experience of Maths Anxiety was quite difficult, as she didn't easily engage with discussions about Maths. Her language of choice, when she did talk about Maths, wasn't strikingly positive, but it wasn't negative either. Using phrases such as *"it's alright,"* or *'I don't like it, but it's fine,"* might indicate a lack of strong emotion towards Maths, but paired with her anxiety towards it, this could be more avoidant in nature. That is, she didn't want to talk about Maths, so engaged as little as possible in the conversation.

It is also possible that an individual can be apathetic towards using and learning Maths, whereby they hold few beliefs about the subject and therefore have neither compellingly positive or negative emotions towards Maths. This apathy means that they are not driven to make decisions or take action, whether positive or negative, usually resulting in the passive learning of Maths, a lack of motivation, and minimum engagement. While someone who is apathetic towards Maths can be mistaken for someone who has a negative attitude towards it, it is inherently different: negativity often sparks action, whereas apathy leads to a distinct lack of action.

Andrew's Experience

Andrew was quite vocal about his feelings towards Maths and would eagerly explain why it was the worst subject of them all. From being too difficult, to being boring, and to having the worst teachers of the lot, he couldn't identify one good thing about Maths. Andrew was excited at the prospect of no longer continuing to study Maths, deeming it a *"waste of time"* nowadays. He couldn't remember how he felt about Maths when he was younger, but he did say that the longer he has been learning Maths, the more he has hated it.

As a general trend in data, attitudes towards Maths have been found to decline as students get older. Longitudinal findings have shown that, often, students begin school with a generally positive attitude towards Maths but leave school with a notably more negative attitude. Whether negative attitudes develop due to bad experiences with Maths, or an individual finds it difficult, boring, or uninspiring, this decline does have an impact on future choices and is a barrier to mathematical learning and progress.

> Popham (2005) found that student attitudes are influential towards mathematical success and are powerful predictors of student behaviour.

Maths Anxiety has been related to the decline in positivity towards Maths. Unsurprisingly, individuals experiencing Maths Anxiety have been found to have negative attitudes towards Maths and often report less enjoyment when using or learning Maths, compared to their less anxious peers. This can negatively impact their engagement with using and learning Maths, as negative attitudes can lead to avoidance and a lack of motivation, two further constructs within the experience of Maths Anxiety discussed in this chapter. This not only highlights the interlinked nature of Maths Anxiety but also demonstrates how influential attitudes are in relation to someone's experience of Maths and Maths Anxiety.

> Hembree (1990) examined correlations between both secondary school students' and college students' enjoyment of Maths and their levels of Maths Anxiety in a meta-analysis. It was reported that they are positively correlated, meaning that students with higher levels of Maths Anxiety reported lower enjoyment of Maths compared to those who were less anxious. In addition, it was found that positive attitudes towards Maths were related to low levels of Maths Anxiety. These correlations were stronger in secondary school students compared to those in post-secondary education.

Correlational research has shown how negative attitudes and high levels of Maths Anxiety are related, but there has been speculation as to whether Maths Anxiety does have a role to play in declining attitudes towards Maths. Does Maths Anxiety lead to this decline in positivity, or does the decline lead to the development of Maths Anxiety? Research can only speculate on the direction of this relationship but is it likely that both Maths Anxiety and negative attitudes fuel each other in a cyclical pattern.

> Tuohilampi et al. (2014) identified a decline in attitudes towards Maths among 3,502 students in Finland. Participants often had a positive attitude before joining primary school, which deteriorated over time until they left school with an increased negative attitude. They also reported that self-belief is likely to decrease during lower secondary school years. However, this wasn't found to impact, or be related to, Maths performance.

As it is more likely that a maths anxious individual will hold a negative attitude towards Maths, identifying the characteristics related to this may also help identify Maths Anxiety itself. While not everyone who experiences Maths Anxiety will have a negative attitude towards Maths, it may be an indicator for teachers to keep in mind, particularly as it has been shown to be related to other aspects of the experience of Maths Anxiety, such as a lack of motivation and avoidance. Arguably, supporting students in developing a positive attitude towards Maths may help address, or at least be a step towards addressing, Maths Anxiety. It could also improve their overall relationship with Maths and potentially their future educational and career choices.

Self-belief

Self-belief is one of the concepts linked to Maths Anxiety which is perhaps the most relatable, as many people have grappled with self-belief at some point in their lives. Lacking the belief in yourself that you can accomplish even the smallest of tasks can significantly influence your self-perception in a negative way. You may have encountered a lack of self-belief in your ability to complete an important task given to you at work, or you may have experienced it when navigating your way through a foreign city to find your hotel. This lack of self-belief may present itself as a recurring issue in one specific situation, or across various scenarios, or emerge only once in a while.

A Note on Terminology

When looking at the research available that focuses on Maths Anxiety and its relationship with self-belief, it is important to note that not all researchers use the phrase "self-belief." The concept of self-belief can also be referred to as "self-rating" (e.g. Dowker, Sarkar and Looi, 2016) or "self-concept" (e.g. Ahmed et al., 2012) and has been used interchangeably with the term "confidence" (e.g. Shores and Shannon, 2007).

However, researchers also often use the term "confidence" in reference to how confident a student is with a solution they have provided in response to a mathematical problem, i.e. posing a question after an answer has been provided, such as *"How confident are you with the answer you have given?"*. This latter question relates to the concept of metacognition, something that has only received a small amount of attention within the field of Maths Anxiety. Please see Table 3.1 in Chapter 3 for a brief description of some of the specific self-belief variables often studied alongside Maths Anxiety.

No matter the terminology used, there is a wealth of research out there that highlights the relationship between Maths Anxiety and an individual's belief in their own ability to use and learn Maths.

A lack of self-belief in an individual's own capacity to use or learn Maths is not uncommon. This has captured the attention of researchers who have explored people's interactions with Maths and their emotional responses to it. An individual having a lack of self-belief in their own mathematical abilities may potentially be related to the perception of the subject's binary nature, where answers are classified as either right or wrong, with no in-between. This can lead to individuals categorising themselves as "bad at Maths," and comparatively, perceiving others to be "good at Maths."

> **Evie's Experience**
>
> A lack of self-belief was apparent in Evie's experience of Maths Anxiety, as she felt that she was never any good at Maths. Having consistently bad marks on Maths tests solidified this belief in her mind, as Evie felt assured that she wouldn't pass her GCSE exam or get into the college she wanted to go to because of failing Maths.

Having a "right or wrong" or "good or bad" perspective of Maths is an oversimplification of the true nature of Maths, and it can be associated with someone developing negative attitudes towards Maths and doubting their ability to achieve. This in turn can negatively impact someone's future mathematical attainment, their perception of Maths itself, their perseverance with the subject, and effort overall when using or learning Maths.

> Nabila and Widjajanti (2020) emphasise the importance of self-belief in the process of learning Maths. They identified that some students do not hold a positive view of their mathematical ability or potential, which can negatively impact their future mathematical achievements.

In a mathematical situation, low self-belief may be presented in various ways. Some individuals may vocalise doubts about their ability to succeed or may refrain from even attempting tasks as they don't feel that they will succeed. Others may describe how they have hit a "mathematical wall" and can't progress any further, or attribute mathematical success to natural, innate talent, which they feel they simply do not have. This fixed mindset can also lead to unfavourable comparisons of themselves to others, where they may constantly view themselves as less capable. This can impact effort, and someone with low self-belief in Maths may not participate in discussions or try their hardest, as they don't feel there is any point if they fail anyway.

Alternatively, someone with low self-belief in Maths could instead overcompensate by putting in more effort and trying extremely hard, as they believe they need to try harder to successfully learn and use Maths. Someone may also fear making mistakes if they have a lack of self-belief, or be overly sensitive to criticism, making it difficult to accept constructive feedback. Sadly, having a lack of self-belief may also impact an individual's choice to continue learning Maths in their education, dropping the subject as soon as they are able to.

It is important to note, however, that someone's lack of self-belief in Maths does not necessarily reflect their actual mathematical ability and attainment. For example, an individual may do really well on a Maths test but doubt their ability to successfully learn or use Maths, despite having objective evidence that they can. Their subjective perception that they are not proficient in Maths may or may not align with their objective attainment, for example, achieving 80% on a Maths test, but this can still have a negative impact regardless and strengthen their belief that they are not "good at Maths."

> **Sarah's Experience**
>
> Let's go back to Sarah's experience of Maths Anxiety. Despite having above average attainment, and doing well in mathematical tasks and testing situations, Sarah had a distinct lack of self-belief in her mathematical abilities. She often used negative self-talk, such as phrases like *"I just can't do it, it's too hard for me."* While this seemingly wasn't related to her ability, she did reference a previous experience as to why she felt that she wasn't mathematically able, where she felt embarrassed by answering a question wrong in class and the teacher *"looked annoyed and shook her head in front of everyone."* She could recall this experience in vivid detail, as well as how it made her feel in that moment. This highlights that self-belief may be integral within the experience of Maths Anxiety, but it doesn't necessarily relate to attainment - it could be related to other people and previous experiences.

When someone believes that they are unskilled in Maths, it has the potential to hinder their attainment and progress. This can foster a recurring cycle of negative self-belief, which has been found to negatively impact mathematical attainment. For example, if someone believes that they are not able to understand Maths and use it successfully, the more they believe this and identify with this narrative, the more likely it is that their mathematical outcomes will suffer and be negatively impacted, such as their attainment. This illustrates the negative pattern of a self-fulfilling prophecy.

> Hannula, Maijala, and Pehkonen (2004) conducted a longitudinal study spanning 18 months to examine the self-belief and mathematical attainment of more than 3,000 students, aged between 10 and 13 years old. Their data indicates that an individual's ability to learn Maths is influenced by their own self-belief about their ability to achieve.

Many researchers have identified that having a lack of self-belief is part of the experience of Maths Anxiety, suggesting that maths anxious individuals often perceive themselves as lacking proficiency in Maths as they feel they are unable to understand it, or use it correctly. While some maths anxious individuals may also contend with a lack of self-belief in other areas of their lives, it is the particular lack of self-belief in their ability to use and learn Maths that is related to their experience of Maths Anxiety.

> Hembree's (1990) meta-analysis of 151 studies found an average correlation of -0.71 between Maths Anxiety and Maths self-concept, evidencing the strong relationship between the two constructs.

Plenty of research suggests that the stronger someone's experience of Maths Anxiety is, the lower their self-belief in their mathematical abilities may be, and vice versa. While there will of course be anomalies and experiences that don't follow this data trend, it is likely that having low-self-belief is an indicator of Maths Anxiety.

To date, there is no existing literature that has specifically delved into the experiences of individuals with both low self-belief and low levels of Maths Anxiety, or those individuals with high self-belief and high levels of Maths Anxiety. That's not to say such profiles don't exist, but they are perhaps less common and more difficult to identify.

> Jain and Dowson (2009) studied 232 students in India (mean age = 13 years) and found that poorer self-regulation (the propensity to set goals, plan strategies, and evaluate and modify ongoing behaviour) was related to higher Maths Anxiety. However, this relationship was mediated (statistically explained) by a student's level of self-efficacy.

While there is agreement that having a lack of self-belief and experiencing Maths Anxiety are connected, there is no agreement on the casual direction of this relationship. It's unclear whether having low self-belief causes Maths Anxiety to develop, or instead, whether experiencing Maths Anxiety subsequently results in low self-belief, as both scenarios are plausible. What is interesting, however, is that some intervention work (as we will explore further in Chapter 6) does not always result in a positive change in both Maths Anxiety and self-belief.

If someone were to initially lack self-belief about their ability to successfully learn and use Maths, this may trigger heightened anxiety in these situations, ultimately leading to the development of Maths Anxiety. Alternatively, if someone is already experiencing Maths Anxiety and is continually confronted with mathematical situations that evoke anxiety, this can negatively impact their belief that they can successfully achieve, leading to diminished self-belief.

> Meece, Wigfield, and Eccles (1990) conducted longitudinal research and found that when students were 11 years old, their perceived ability in Maths predicted both Maths Anxiety and how well they thought they would do in Maths when they were 13 years old. If students had low self-belief, they were more likely to develop higher Maths Anxiety two years later, suggesting that self-belief leads to Maths Anxiety. Interestingly, students' perceptions of their Maths ability mediated the relationship between Maths Anxiety and previous Maths performance. This suggests that it's a student's interpretation of their Maths achievement, not the actual outcome or grades, that has the strongest effect on a student's Maths Anxiety.

The majority of research, especially qualitative accounts from people, suggests that low self-belief leads to the development of Maths Anxiety, but there is a growing body of evidence that suggests that both Maths Anxiety and self-belief can influence each other reciprocally. When a maths anxious individual encounters a mathematical situation where they feel that they cannot succeed, this may lead to a decline in their self-belief. This diminished self-belief can in turn intensify an individual's Maths Anxiety, as their anxiety in mathematical scenarios may escalate because they don't believe they can achieve, causing discomfort and an increased perception of failure and lack of self-belief. This negative cycle of heightened Maths Anxiety and low self-belief can persist, leading to negative experiences when using and learning Maths.

> Ahmed et al. (2012) researched the relationship between Maths Anxiety and self-belief (in this case, Maths self-concept) over a longitudinal study lasting three years, to see if it was reciprocal; did Maths Anxiety affect self-belief, and did self-belief also affect Maths Anxiety? In a sample of 495 10 and 11 years old, they reported that both Maths Anxiety and self-belief had an impact on each other over time, but self-belief had a stronger impact on Maths Anxiety three years later. This research was correlational in design, making it difficult to confirm casuality, but the use of data over a three-year period strengthens their claim.

This leads to the consideration of whether a lack of self-belief can serve as an early predictor, or indicator, of Maths Anxiety developing later on. If this were the case, it would be valuable for teachers to keep this in mind when working with students in the classroom. Identifying students with low self-belief in Maths could lead to subsequent support if it appears that Maths Anxiety has also developed. Early action to address poor self-belief might even mitigate the development of Maths Anxiety in the first place, and work to foster a positive classroom environment.

> Johnston-Wilder, Lee, and Mackrell (2021) highlight the way in which Self-Determination Theory (Ryan and Deci, 2018) can be used to explain how Mathematical Resilience can be used against Maths Anxiety. Having its roots in positive psychology, this theory is based on the premise that people have agency and a natural tendency towards growth. To facilitate and ensure Maths students experience psychological safety and well-being, three basic psychological needs should be met:
>
> **COMPETENCE:** to minimise feelings of failure and inadequacy, a student's need for competency should be met. They should have the means to improve mathematically.
> **RELATEDNESS:** to avoid feeling isolated and abandoned, Maths learning contexts should nurture a sense of belonging. Students should feel valued and connected.

> **AUTONOMY:** this is where a Maths student feels their actions are volitional and not based on internal or external pressure. Regulation of actions should be in accordance with authentic interests and values. Accordingly, when a student is autonomous in their self-regulation, they are self-initiating and persistent. This is because the tasks they engage in are perceived as interesting or personally important to them.

Motivation

All actions we take are driven by our motivation, as it pushes us to set goals and take action to achieve them. This might be something that happens on a daily basis, such as waking up without snoozing your alarm so that you get to work on time, or it could be a larger, long-term goal where you need sustained motivation to achieve it, such as a change in career or buying a house. Goals can be short-term or long-term, big or small, but they all require motivation to be achieved. If we lack motivation, very little action will be taken as there is no driving force pushing us to succeed. Having a lack of motivation can impact small decisions and actions, as well as big ones, due to our personal beliefs and desires.

No matter the goal, motivation is needed to work towards achieving something that is important to us and will likely develop from the desire for a sense of purpose or a reward. Often, motivation is categorised as either intrinsic or extrinsic. Intrinsic motivation refers to when someone is driven to succeed in their goal because of the enjoyment and satisfaction they feel. For example, you might be motivated to learn new skills related to your job because you really enjoy learning something new and being able to share this with your colleagues. You love the feeling of satisfaction when you accomplish these skills and enjoy being told by others how impressed they are with your skill development. These comments have a positive impact on you and increase your intrinsic motivation to continue learning new skills in the future.

Alternatively, extrinsic motivation refers to when someone is driven to accomplish something due to the incentive of being rewarded, or avoiding punishment. This involves a sense of gain, and receiving something positive, or evading something negative. Following on from the earlier example, you could be extrinsically motivated to learn new skills that are relevant to your job in order to gain a promotion. Alternatively, if you are aware of redundancies at work, you may be extrinsically motivated to learn new skills and become better at your job to avoid redundancy.

> Zakaria and Nordin (2008) found that Maths Anxiety was negatively correlated with achievement, and also effectance motivation – that is, whether students enjoy or seek challenges in Maths.

Having the motivation to learn and use Maths varies widely. Some students are intrinsically motivated due to the sense of achievement and enjoyment they have from being mathematically successful. Other students may be extrinsically motivated to successfully use Maths as they strive for the certificates and physical rewards associated with mathematical success. Regardless of the type of motivation, research has shown that students who are highly motivated to succeed in Maths do have higher attainment and a better understanding of the application of Maths, compared to students who are less motivated. They are also more likely to enjoy Maths lessons and have the desire to participate in discussions and class activities.

> Aunola, Leskinen, and Nurmi (2006) reported a developmental cycle between motivation and mathematical attainment, whereby high attainment increased subsequent motivation towards Maths, which then led to predicted higher performance in the future.

On the other hand, student motivation can be quite low when engaging with Maths. Some students have very little drive to want to learn and use Maths successfully, as they don't have the desire to involve themselves with the subject or develop their mathematical knowledge. Their low motivation aligns with having a lack of enjoyment and satisfaction associated with Maths, and a lack of appreciation of any physical rewards associated with success, which might further lower their motivation.

Some students may remain consistently unmotivated in Maths, yet some will always remain highly motivated. Motivation is not often stagnant though. As teachers, we are aware of the impact daily life can have on the motivation to learn and use Maths, and it is likely that for most students, their motivation will ebb and flow according to their current priorities and personal desires. This could be related to the mathematical content they are learning, or even a lack of sleep, for example.

> Spinath, Freudenthaler, and Neubauer (2010) found that students' interest in Maths was significantly positively correlated with Maths attainment. However, the correlation was quite weak, and even weaker for girls compared to boys. It is also open to debate as to what "interest" in Maths really involves.

A student's level of motivation to learn and use Maths is likely to be related to a range of factors. Research has focused primarily on the relationship between attainment and motivation. If a student does well in assessments and has high attainment, they are likely to be motivated to continue learning, and engaging with, Maths, while those with low attainment are likely to feel demotivated and engage with the subject less. Attainment can actually negatively impact motivation, with research finding that motivation levels decrease throughout students' educational years if they find learning Maths challenging. Interlinked with attainment is an individual's self-belief, which has also been found to impact levels of motivation. If a student has a high level of self-belief in their ability to achieve mathematically, their motivation is also likely to be high, while those who have low self-belief are likely to be less motivated.

This can create a negative cycle, where low attainment, low self-belief, and low motivation can feed into negative attitudes towards Maths, and a lack of enjoyment of Maths and appreciation of its usefulness. If someone genuinely enjoys learning and using Maths, believes they are capable of doing so, and has high attainment, their motivation is likely to be high. If someone has low attainment, doesn't believe they can achieve in the subject, and doesn't enjoy learning about it, they are likely to have low motivation. This shows how interlinked motivation to learn and use Maths is with other factors, and how important having high motivation levels can be upon a student's relationship with Maths, their attitude towards it, and their attainment.

As discussed in this chapter so far, self-belief, attainment, and negative attitudes are all related to Maths Anxiety, so it is not surprising that motivation has been linked to the experience as well. It is unclear whether a lack of motivation impacts the experience of Maths Anxiety, or whether Maths Anxiety can lead to a lack of motivation, or both, but researchers have consistently found a relationship between levels of motivation and Maths Anxiety. Specifically, many researchers have identified that a maths anxious individual is likely to have lower motivation to learn and use Maths compared to students who are less anxious.

This may reflect how having negative attitudes towards Maths, not enjoying Maths lessons, having a lack of self-belief, or low attainment as part of their experience of Maths Anxiety may link to being unmotivated in relation to Maths as well. Instead, a maths anxious individual may be more motivated to avoid situations involving Maths, rather than participate. While avoidance of Maths will be discussed later in this chapter, it is possible to empathise with a student who appears unmotivated if it means they avoid engaging with the very thing that reinforces their negative self-belief and makes them feel anxious.

> Hembree (1990) found that students with high levels of Maths Anxiety had lower motivation to learn Maths, with a mean, moderately strong, negative correlation (r) of 0.64.

Some students may be unmotivated when it comes to learning in general, regardless of the subject, but it is possible that maths anxious students' lack of motivation may only be apparent in Maths. They may be engaged in other subjects across the curriculum, evidencing how their anxiety, and lack of motivation, is specific to Maths. A lack of motivation may be seen through passive behaviour, such as not engaging with discussions of activities in Maths lessons, or avoiding mathematical situations. In addition to this, a maths anxious student may also show a lack of motivation through their body language and posture, such as slouching or having their head in their hands.

> **Aki's Experience**
>
> Aki was highly demotivated within his Maths lessons, which contrasted to how he was within other lessons. His teacher shared that it was so difficult to enthuse Aki in Maths lessons, as he was always slumped in his chair and looking out the window, rather than being involved. This created a sense of frustration in his teacher, as she noted that in all other lessons, he was much more focused and part of the classroom, particularly in Science and History. Clearly, Aki's lack of motivation was only apparent in Maths within the classroom, which was where he felt anxious.

Feeling of overwhelmedness is also something that could identify low motivation in Maths lessons, and procrastination may occur regularly. It is likely that a maths anxious student with low motivation may just want the lesson to finish, rather than having a focus on developing their understanding. While this will differ among maths anxious students, looking out for these behaviours and indicators of low motivation will support teachers in positively influencing Maths Anxiety and a lack of motivation in the classroom, leading to a more positive relationship with Maths overall.

We should also consider the relevance of effort in a bit more detail here. While there might be an assumption that low effort accompanies a lack of motivation, the situation might become complicated when we also introduce Maths Anxiety. A maths anxious individual may feel safer by avoiding Maths, and therefore not put effort into learning Maths. This is likely to lead to a lack of mathematical understanding and progression and hinder their future career choices. Alternatively, by putting effort into learning and using Maths, they are placing themselves in the middle of what makes them anxious, and if they do not succeed in what they aimed to accomplish, this can perpetuate feelings of anxiety, and negative perceptions of themselves, and Maths.

> Skemp (1971) discussed the importance of effort when considering Maths Anxiety, whereby, for the maths anxious student, effort might become a no-win situation. A lack of effort, perhaps displayed as avoidance, is likely to have a detrimental impact on mathematical learning. However, investing effort is also a risk for a maths anxious student; they might already be fearful of failure, and subsequent perceived failure following invested effort might only serve to reinforce existing negative self-beliefs and exacerbate their Maths Anxiety.

From this perspective, effort becomes a rather unhelpful aspect of Maths Anxiety. On top of this, we should be mindful of the hidden effort that sometimes takes place. Research into Maths Anxiety and Maths performance tends to show a more consistent relationship exists between Maths Anxiety and how long it takes a person to complete a Maths task, compared to the relationship between Maths Anxiety and how accurate people are in their answers. This links back to what we discussed earlier about processing efficiency. Perhaps, Maths Anxiety impacts cognitive processing in a way that means the maths anxious student has to exert greater effort to meet the demands of the Maths task. What that effort relates to exactly is open to debate, but it almost certainly involves attentional processes, working memory, and having to deal with unhelpful, worrisome thoughts.

> **Marc's Experience**
>
> When Marc was reflecting on the week's Maths lessons, he felt that he had tried really hard. He noted lots of things he found difficult, such as staying focused, asking for help when he found things hard, and completing the task in the time given. He was adamant that he had tried his best. This is what made him feel defeated when the teacher often made comments to *"speed up,"* or that he *"could have done more than that,"* because he genuinely felt that he was putting in as much effort as he could. Marc shared that this did make him feel nervous each lesson, because he felt that he wasn't good enough to meet the expectations of the teacher.

This is a complex area and more research is needed that focuses on the interrelation between Maths Anxiety, motivation, and Maths performance, especially in terms of changes over time. It is likely that these variables interact in a range of ways. Therefore, when supporting maths anxious students, we need to consider the complexities surrounding their level of motivation, and perceived effort, and how this may relate to other areas within their experience of Maths Anxiety.

> Wang et al. (2015) noted that the relationship between Maths Anxiety and performance is not linear when different levels of motivation are considered. In particular, when motivation was high and Maths Anxiety was modest, students' performance was at its highest. However, when Maths Anxiety got past a certain point, performance rapidly dropped even when motivation was high.

Avoidance

By keeping away from something, or preventing something from happening, avoidance is a behaviour which allows someone to steer clear from particular situations, places, or activities that they perceive as threatening or unpleasant. These situations can evoke anxiety, stress, or fear, so avoidance can be seen as a coping mechanism to prevent this from happening. We tend to avoid the things that make us anxious, and there are many examples of this. For instance, if someone is afraid of dogs, they might be more likely to cross the street if they see a dog coming their way. In many ways, Maths Anxiety is not much different to this.

Unlike many situations where we can choose to either engage with, or avoid the thing we are afraid of, Maths is genuinely all around us. Whether splitting the bill at the end of a meal, or following a recipe to bake a cake, Maths pervades so many aspects of our lives that it is almost impossible to avoid it completely. While technology and online search engines are often used to provide mathematical assistance where needed, to convert currency rates while on holiday, or calculate the too-good-to-be-true discount when shopping, for example, these instances can still spark Maths Anxiety and initiate avoidance of the situation.

While people who are not maths anxious may also avoid mathematical situations, research has indicated that avoidance is more highly associated with Maths Anxiety. Avoidance of Maths is an interesting behaviour associated with Maths Anxiety, as learning and using Maths is avoided primarily due to the anxiety it can evoke, rather than necessarily being due to a lack of enjoyment or willingness to learn. The principles of Behaviourism tell us that we are quite proficient in making links between external events and how we feel, as a response to those events. So, if something we are doing, or something that is happening to us, makes us feel anxious, it is likely that we will remove ourselves from that situation and avoid similar situations as best we can in the future.

> Ashcraft and Krause (2007) highlighted a global avoidance pattern, suggesting that Maths Anxiety often leads to heightened avoidance of mathematical situations, including future career paths related to Maths.

In the case of Maths Anxiety, a person may begin to associate negative feelings with learning or using Maths, increasing their feelings of anxiety in mathematical situations, and therefore become likely to avoid these to prevent experiencing Maths Anxiety. This might happen for all sorts of reasons, such as someone feeling embarrassed if they struggled to answer a Maths question posed by a teacher in front of peers, or they may experience panic when completing a Maths test because they think they can't do it. Although only a couple of examples, it follows that future mathematical situations become something to avoid.

> Butler (1998) found that 10-12 years old avoided asking for help when needed in Maths lessons for a variety of reasons: they wanted to figure things out on their own, they were worried that asking for help would show they found Maths difficult, or they thought it wouldn't make a difference to their understanding. Students, especially boys, avoided asking for help mostly because they were worried about seeming "bad at Maths."

Avoidance can be seen as a way in which a maths anxious individual copes with being presented with a mathematical situation that negatively affects them. It is an ongoing battle for adults who experience Maths Anxiety, as mathematical situations are not always avoidable, such as calculating the amount of time you have before setting off to work, or working out how much it is going to cost to fill your car with petrol. However, in situations where others can engage with the imposed use of Maths on their behalf, a maths anxious individual is likely to avoid any involvement and allow other people to complete the task for them. For example, they may rely on others organising their monthly outgoings, or they might even pay for a group's meal in a restaurant just to avoid the stress of calculating everyone's share. Any steps taken to avoid using Maths may be seen as positive in the eyes of someone who is experiencing Maths Anxiety.

Maths avoidance is much harder for those maths anxious individuals who are still at school, and learning and using Maths on a daily basis. In the classroom, maths anxious students may take active steps to avoid learning or using Maths through behaviours such as avoiding their teacher's questions, and eye contact, or not putting their hand up to answer questions. They may avoid self-assessing their work due to not wanting to know the outcome, avoid completing the lesson's activity altogether, or physically remove themselves from the classroom by either not attending in the first place or going to the toilet on multiple occasions throughout a lesson.

A maths anxious student may instead be passively avoidant, whereby they appear disengaged during their Maths lessons and do not join in any activities, discussions, or tasks, whether this is in pairs, groups, or class settings. This is still purposeful, as it is like being a passive bystander. Maths is being discussed, learnt, and used all around them, but they are choosing not to be a part of it. A maths anxious student can be actively avoidant of learning and using Maths, or passively avoidant, or even display behaviours associated with both. All forms of avoidance lead to them keeping Maths at arm's length.

Phoebe's Experience

During each Maths lesson where Phoebe was observed, she showed avoidant behaviours. These ranged from avoiding eye contact when the teacher was asking questions, to sitting back in her seat and staring out the window when the class were discussing a mathematical concept. When working, she didn't engage in mathematical discussion with her classmates, and when it appeared that she was finding something challenging, she did not ask for help. As soon as the bell rang at the end of each lesson, she left.

At this point, it is worth discussing the concept of intentional versus unintentional avoidance. When we hear the term "avoidance," we tend to think of it as intentional; that students purposely go out of their way to avoid placing themselves in a particular situation, like active avoidance, or purposefully avoid engaging with a task of some kind, like passive avoidance. Indeed, as discussed in the case of Maths Anxiety, this is often seen.

However, some of the behaviours we see, and some of the accounts we hear, tend to suggest that people do not always intentionally avoid Maths. For example, unintentional avoidance might come in the form of procrastination, such as putting off doing Maths homework, or rushing through tasks, to simply get Maths homework done and dusted. While these behaviours could be intentional, a student who is highly maths anxious might not necessarily think of them as avoidance per se. Rather, they might have other intentions, such as prioritising homework for other subjects, or simply aiming to complete their Maths work as quickly as possible.

> ### Dan's Experience
>
> When completing homework, Dan always rushed through it. He said that he wasn't bothered about it being right or wrong, it just *"had to be done."* He said that nothing ever bad came from getting things wrong, his teacher expected it in fact, but if he didn't do it, that's when there were consequences. Dan explained that he rushed through it because he hated it, didn't understand it, and didn't see the point in it.

In a similar vein, after several years of Maths learning and testing, a highly maths anxious student might come to almost automatically disregard, or even fail to notice, Maths learning opportunities. Taking the example of dogs, if a person is anxious around dogs, they are unlikely to pay much attention to them, perhaps other than to initially recognise an animal as a dog and to then go out of their way to avoid it; they are unlikely to show any interest, whether consciously or subconsciously, in learning about dogs. Applying this analogy to Maths, a maths anxious student might come to dismiss Maths, to almost separate themselves from it. This form of avoidance can be particularly tricky to address, especially as it is likely to be experienced by students whose Maths self-belief is low.

Perhaps what is most interesting is the notion that anxiety can affect attentional processes at very early stages in the processing of mathematical information. This suggests that high levels of Maths Anxiety, in some way, may operate below the level of conscious awareness. However, a question remains over the extent to which such early attentional processes indicate the unintentional avoidance of mathematical stimuli. Future work utilising eye-tracking technology is likely to offer some answers to this, but at face value, it is easy to conceptualise how someone might come to avoid paying attention to Maths without even realising it.

In the short term, for someone who is anxious about Maths, avoidance is likely to offer at least some degree of temporary comfort. Of course, the problem with this is that associations are further made between avoidance of Maths and the avoidance of negative feelings:

"If I avoid Maths, I avoid feeling anxious, because Maths gives me anxiety."

From this, it is not too hard to see how short-term avoidance of Maths can lead to longer term avoidance. Yet, avoiding Maths does not necessarily stop feelings of anxiety. The anxiety remains somewhat hidden under the surface; it's just not being triggered. When a maths anxious individual is finally presented with Maths again, which is likely, the negative associations they have developed about learning and using Maths may well come to the fore.

Failure to engage with repeat behaviours, whether that is retrieval of multiplication facts or working out a budget for example, can result in feelings of "rustiness" and general low self-belief. This is sometimes experienced in people who have not engaged in formal Maths education for a number of years. In particular, adults who are looking to get back into education after a number of years will often say their Maths knowledge feels "rusty." It might also be the case that mature students are presented with newer ways of learning Maths, which presents challenges around interference with what they are more familiar with. In fact, it should not be underestimated just how many adults might be put off from re-entering education because of the real or perceived challenging Maths content. This can include those who want to do the university degree that they never did when they were younger, and it can include people wanting to embark on specific careers that require evidence of a certain level of mathematical knowledge and understanding. A good example is Nursing, in which Maths Anxiety has been shown to exist. Entrance exams, the Maths involved in medicine management, and the general fear of getting calculations wrong in life-or-death situations, can all create anxiety for students in Nursing and allied healthcare. Maths Anxiety in this context has an impact on recruitment, progression, and retention on taught programmes and apprenticeships.

> Ashcraft (2002) highlighted how avoiding mathematical situations can lead to weakened mathematical competency, but those who are high in Maths Anxiety learn less of the Maths they are exposed to.

In the long term, the avoidance of learning and using Maths can lead to poor mathematical knowledge and attainment. This may occur due to missed opportunities in retrieving and consolidating prior knowledge, as well as learning new mathematical concepts and processes. In the classroom, this may mean that students avoid opportunities to strengthen their previous learning through recall and consolidation, and when learning something new, this is interrupted by avoidant behaviour so it cannot be learnt effectively. It is likely that the more that mathematical learning is avoided, the bigger the gap will become between an individual's attainment and their age-related expected attainment.

In addition to this, several empirical studies have shown how Maths Anxiety is associated with an unwillingness to study Maths in the future, and the intention to avoid careers that are perceived to involve Maths. Worryingly, while this has been found in older students and adults, it has also been found in primary school children. Moreover, longitudinal findings have shown that actual career choice in adulthood can be predicted by Maths Anxiety during adolescence. This shows how powerful Maths Anxiety can be, along with the urgent need to address it early on.

> Ahmed (2018) found that Maths Anxiety during adolescence predicted STEM career choice approximately 20 years later. Specifically, adolescents who were categorised as being consistently low in Maths Anxiety during adolescence were approximately 7.4 times as likely to be employed in a STEM career compared to those who had been categorised as being consistently high in Maths Anxiety. Similarly, those who were recorded as having a decreasing level of Maths Anxiety during adolescence 20 years earlier were 6.4 times as likely to be employed in STEM as adolescents categorised as being consistently high in Maths Anxiety.

As research has shown that Maths Anxiety and avoidance are linked, it is likely that by identifying the patterns and behaviour of avoidance within the classroom, teachers may be able to identify maths anxious individuals and work to support them. Of course, as discussed before, a student who is not experiencing Maths Anxiety can also avoid learning and using Maths. So, while it is not a guaranteed way in which to successfully identify Maths Anxiety, it could be a useful indicator, and a platform from which to address students' Maths Anxiety and reduce the occurrences of avoidant behaviour. This may also have a positive impact on students' mathematical attainment, as more learning may occur in the absence of avoidance.

So... What's Involved?

- There are multiple factors involved in the experience of Maths Anxiety, which can differ for each individual.
- Low attainment is often associated with the experience of Maths Anxiety.
- An individual's working memory can be negatively impaired by Maths Anxiety.
- Negative Maths attitudes are often part of the experience of Maths Anxiety.
- Maths anxious individuals are likely to have a lack of self-belief in their mathematical capabilities.
- A lack of motivation is often shown by maths anxious individuals, but sometimes high levels of motivation and Maths Anxiety can exist together.
- Avoidance is a key behaviour that can characterise Maths Anxiety.
- Establishing causal relationships between Maths Anxiety and factors such as attitudes, motivation, self-beliefs, and avoidance is difficult. In reality, such relationships are likely to be bidirectional and not always linear.
- Being aware of these factors may help us to identify Maths Anxiety in the classroom, leading to support for maths anxious students.

There are clearly a multitude of factors involved in the experience of Maths Anxiety. Based on a wealth of research, it can be suggested that a maths anxious individual is likely to have low attainment and find learning and using Maths difficult. This is interlinked with the role of the working memory, which anxiety can negatively impact. Having low attainment and impaired working memory can feed into having low self-belief, as maths anxious individuals doubt their mathematical capabilities and feel that they cannot succeed. They may also have a negative attitude towards Maths, believing it to be too difficult, boring, or uninspiring. This may relate to having a lack of motivation, as a maths anxious individual may not see the value in engaging with Maths, preferring to avoid the subject altogether instead.

While these factors are constructs in their own right, they overlap and impact each other in a way that can make it difficult to sustain a positive relationship and perception of Maths. It is also important to note that within each experience of Maths Anxiety, these factors may not present themselves in the same way, or to the same extent. For example, having a lack of self-belief may be the most significant factor in some maths anxious individuals' experiences, whereas for others, it may play a minor role, or not be a factor at all. Some maths anxious individuals may experience a negative cycle relating to having low mathematical attainment, high levels of anxiety, and a lack of motivation, while another maths anxious individual may have an extremely detrimental lack of self-belief feeding into their Maths Anxiety in a cyclical manner.

The experience of Maths Anxiety is truly unique to each individual, and while these factors have been found to commonly occur across experiences, how they are seen and presented may differ as well. Therefore, it's important to bear in mind that Maths Anxiety is complex and involves lots of different factors that may or may not be consistently present. However, knowing some commonalities across experiences of Maths Anxiety can be useful in identifying maths anxious individuals and supporting them to develop a more positive relationship with Maths. By truly understanding what is involved in an individual's experience of Maths Anxiety, we are more equipped in our knowledge of how to best help them reduce their feelings of anxiety and work towards improving their unique experience through a targeted approach.

Later in the book, we will look at how understanding each individual's experience can inform purposeful and directed actions and best support those experiencing Maths Anxiety. The next chapter will consider who, and what, can influence Maths Anxiety. Before then, we have provided a conceptual framework, depicted below in Figure 4.1. This is our attempt to capture what appear to be the primary factors involved in Maths Anxiety, from a broad perspective. It is by no means exhaustive, but it provides a visual aid for understanding the range and complexity of what is involved. The directions of the arrows indicate the main way in which each factor is generally accepted as predicting, and potentially causing, another. It is likely that there are more connections between factors, including bidirectional ones, than we have included here. Nevertheless, the framework gives an overview of the types of factors involved in Maths Anxiety, including many individual differences and external factors. We hope this is a useful starting point for anyone trying to get to grips with the multitude of factors associated with Maths Anxiety.

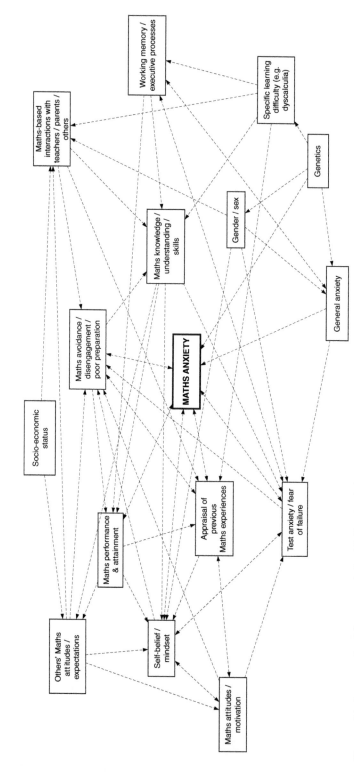

Figure 4.1 A conceptual framework for Maths Anxiety.

References

Ahmed, W. (2018) 'Developmental Trajectories of Math Anxiety during Adolescence: Associations with STEM Career Choice', *Journal of Adolescence*, 67, pp. 158-166. Available at: https://doi.org/10.1016/j.adolescence.2018.06.010.

Ahmed, W., Minnaert, A., Kuyper, H. and Van der Werf, G. (2012) 'Reciprocal Relationships Between Math Self-Concept and Math Anxiety', *Learning and Individual Differences*, 22(3), pp. 385-389. Available at: https://doi.org/10.1016/j.lindif.2011.12.004.

Ashcraft, M. H. (2002) 'Math Anxiety: Personal, Educational, and Cognitive Consequences', *Current Directions in Psychological Science*, 11(5), pp. 181-185. Available at: https://doi.org/10.1111/1467-8721.00196.

Ashcraft, M. H. and Kirk, P. E. (2001) 'The Relationships Among Working Memory, Math Anxiety, and Performance', *Journal of Experimental Psychology: General*, 130, pp. 224-237. Available at: https://dx.doi.org/10.1037//0096-3445.130.2.224.

Ashcraft, M. H. and Krause, J. A. (2007) 'Working Memory, Math Performance, and Math Anxiety', *Psychonomic Bulletin & Review*, 14, pp. 243-248. Available at: https://doi.org/10.3758/BF03194059.

Ashcraft, M. H. and Moore, A. M. (2009) 'Mathematics Anxiety and the Affective Drop in Performance', *Journal of Psychoeducational Assessment*, 27(3), pp. 197-205. Available at: https://doi.org/10.1177/0734282908330580.

Aunola, K., Leskinen, E. and Nurmi, J. E. (2006) 'Developmental Dynamics Between Mathematical Performance, Task Motivation, and Teachers' Goals During the Transition to Primary School', *British Journal of Educational Psychology*, 76(1), pp. 21-40. Available at: https://doi.org/10.1348/000709905x51608.

Baddeley, A. D. and Hitch, G. J. (1974) 'Working Memory', in Bower, G. H. (ed.) *The Psychology of Learning and Motivation*. New York: Academic Press, pp. 47-89. Available at: https://doi.org/10.1016/S0079-7421(08)60452-1.

Butler, R. (1998) 'Determinants of Help Seeking: Relations Between Perceived Reasons for Classroom Help-Avoidance and Help-Seeking Behaviors in an Experimental Context', *Journal of Educational Psychology*, 90(4), p. 630. Available at: https://doi.org/10.1037/0022-0663.90.4.630.

Carey, E., Hill, F., Devine, A. and Szűcs, D. (2016) 'The Chicken or the Egg? The Direction of the Relationship Between Mathematics Anxiety and Mathematics Performance', *Frontiers in Psychology*, 6, pp. 1-6. Available at: https://doi.org/10.3389/fpsyg.2015.01987.

Cipora, K., Rossi, S., Connolly, H., von Bergen, A., Baumgartner, V., Kapur, M. and Gashaj, V. (2023) *Maths Anxiety and Positive Attitudes Towards Mathematics Are Not Mutually Exclusive*. Thessaloniki: EARLI.

Dowker, A., Sarkar, A. and Looi, C. Y. (2016) 'Mathematics Anxiety: What Have We Learned in 60 Years?', *Frontiers in Psychology*, 7, pp. 1-17. Available at: https://doi.org/10.3389/fpsyg.2016.00508.

Eysenck, M. W. and Calvo, M. G. (1992) 'Anxiety and Performance: The Processing Efficiency Theory', *Cognition and Emotion*, 6(6), pp. 409-434. Available at: https://doi.org/10.1080/02699939208409696.

Eysenck, M. W., Derakshan, N., Santos, R. and Calvo, M. G. (2007) 'Anxiety and Cognitive Performance: Attentional Control Theory', *Emotion*, 7(2), p. 336. Available at: https://doi.org/10.1037/1528-3542.7.2.336.

Faust, M. W., Ashcraft, M. H. and Fleck, D. E. (1996) 'Mathematics Anxiety Effects in Simple and Complex Addition', *Mathematical Cognition*, 2(1), pp. 25-62. Available at: https://dx.doi.org/10.1080/135467996387534.

Friso-Van den Bos, I., Van der Ven, S. H., Kroesbergen, E. H. and Van Luit, J. E. (2013) 'Working Memory and Mathematics in Primary School Children: A Meta-Analysis', *Educational Research Review*, 10, pp. 29-44. Available at: https://dx.doi.org/10.1016/j.edurev.2013.05.003.

Hannula, M. S., Maijala, H. and Pehkonen, E. (2004) 'Development of Understanding and Self-Confidence in Mathematics; Grades 5-8', in Høines, M. J. and Fuglestad, A. B. (eds.) Proceedings of the 28th Conference of the International Group for the Psychology of Mathematics Education. Norway: PME, pp. 17-24.

Hembree, R. (1990) 'The Nature, Effects, and Relief of Mathematics Anxiety', *Journal for Research in Mathematics Education*, 21(1), pp. 33-46. Available at: https://doi.org/10.2307/749455.

Hunt, T., Clark-Carter, D. and Sheffield, D. (2014) 'Exploring the Relationship Between Mathematics Anxiety and Performance: The Role of Intrusive Thoughts', *Journal of Education, Psychology and Social Sciences*, 2, pp. 69-75.

Jain, S. and Dowson, M. (2009) 'Mathematics Anxiety as a Function of Multidimensional Self-Regulation and Self-Efficacy', *Contemporary Educational Psychology*, 34(3), pp. 240-249. Available at: https://psycnet.apa.org/doi/10.1016/j.cedpsych.2009.05.004.

Johnston-Wilder, S., Lee, C. and Mackrell, K. (2021) 'Addressing Mathematics Anxiety Through Developing Resilience: Building on Self-Determination Theory', *Creative Education*, 12, pp. 2098-2115. Available at: https://dx. doi.org/10.4236/ce.2021.129161.

Maloney, E. A., Ansari, D. and Fugelsang, J. A. (2011) 'The Effect of Mathematics Anxiety on the Processing of Numerical Magnitude', *Quarterly Journal of Experimental Psychology*, 64, pp. 10-16. Available at: https://doi.org/10.1080/17470218.2010.533278.

Meece, J. L., Wigfield, A. and Eccles, J. S. (1990) 'Predictors of Math Anxiety and Its Influence on Young Adolescents' Course Enrollment Intentions and Performance in Mathematics', *Journal of Educational Psychology*, 82(1), p.60. Available at: https://doi.org/10.1037/0022-0663.82.1.60.

Nabila, L. A. and Widjajanti, D. B. (2020) 'Self-Esteem in Mathematics Learning: How to Develop it Through Contextual Teaching and Learning Approach?', Journal of Physics: Conference Series, 1581(1), p. 12049. Available at: https://dx.doi.org/10.1088/1742-6596/1581/1/012049.

National Mathematics Advisory Panel (NMAP) (2008) Foundations for Success: The Final Report of the National Mathematics Advisory Panel. Available at: https://files.eric.ed.gov/fulltext/ED500486.pdf (Accessed: 22nd April 2024).

Pekrun, R. (2006) 'The Control-Value Theory of Achievement Emotions: Assumptions, Corollaries, and Implications for Educational Research and Practice', *Educational Psychology Review*, 18(4), pp. 315-341. Available at: https://doi.org/10.1007/s10648-006-9029-9.

Popham, W. (2005) 'Students' Attitudes Count', *Educational Leadership*, 62(5), p. 84.

Ramirez, G., Gunderson, E. A., Levine, S. C. and Beilock, S. L. (2013) 'Math Anxiety, Working Memory, and Math Achievement in Early Elementary School', *Journal of Cognition and Development*, 14(2), pp. 187-202. Available at: https://doi.org/10.1080/15248372.2012.664593.

Rossi, S., Xenidou-Dervou, I. and Cipora, K. (2023) 'Emotions and Mathematics: Anxiety Profiles and Their Influence on Arithmetic Performance in University Students', *Royal Society Open Science*, 10(10), pp. 1-22. Available at: https://dx.doi.org/10.1098/rsos.230861.

Ryan, R. and Deci, E. (2018) Self-Determination Theory: Basic Psychological Needs in Motivation, Development, and Wellness. New York: Guilford Press.

Shores, M. L. and Shannon, D. M. (2007) 'The Effects of Self-Regulation, Motivation, Anxiety, and Attributions on Mathematics Achievement for Fifth and Sixth Grade Students', *School Science and Mathematics*, 107(6), pp. 225-236. Available at: https://dx.doi.org/10.1111/j.1949-8594.2007.tb18284.x.

Skemp, R. R. (1971) The Psychology of Learning Maths. London: Penguin. Available at: https://doi.org/10.4324/9780203396391.

Spinath, B., Freudenthaler, H. H. and Neubauer, A. C. (2010) 'Domain-Specific School Achievement in Boys and Girls as Predicted by Intelligence, Personality and Motivation', *Personality and Individual Differences*, 48, pp. 481-486. Available at: https://dx.doi.org/10.1016/j.paid.2009.11.028.

Stoehr, K. J. (2017) 'Mathematics Anxiety: One Size Does Not Fit All', *Journal of Teacher Education*, 68(1), pp. 69-84. Available at: https://doi.org/10.1177/0022487116676316.

Tuohilampi, L., Hannula, M. S., Laine, A. and Metsämuuronen, J. (2014) 'Examining Mathematics-Related Affect and its Development During Comprehensive School Year in Finland', in Nicol, C., Oesterle, S., Liljedahl, P. and Allan, D. (eds.) Proceedings of the 38th Conference of the International Group for the Psychology of Mathematics Education. Vancouver: PME, pp. 281-288.

Vukovic, R. K., Kieffer, M. J., Bailey, S. P. and Harari, R. R. (2013) 'Mathematics Anxiety in Young Children: Concurrent and Longitudinal Associations with Mathematical Performance', *Contemporary Educational Psychology*, 38(1), pp. 1-10. Available at: https://doi.org/10.1016/j.cedpsych.2012.09.001.

Wang, Z., Lukowski, S. L., Hart, S. A., Lyons, I. M., Thompson, L. A., Kovas, Y., Mazzocco, M. M., Plomin, R. and Petrill, S. A. (2015) 'Is Math Anxiety Always Bad for Math Learning? The Role of Math Motivation', *Psychological Science*, 26(12), pp. 1863-1876. Available at: https://doi.org/10.1177/0956797615602471.

Zakaria, E. and Nordin, N. M. (2008) 'The Effects of Mathematics Anxiety on Matriculation Students as Related to Motivation and Achievement', *Eurasia Journal of Mathematics, Science and Technology Education*, 4(1), pp. 27-30. Available at: https://doi.org/10.12973/ejmste/75303.

5 Who and What Can Influence It?

The last chapter highlighted lots of different factors associated with Maths Anxiety, and some of these appear to have a reciprocal effect on it. There are other aspects related to using and learning Maths though, which are not a result of Maths Anxiety but can negatively influence it.

Different things can impact not only the development of Maths Anxiety, but also the intensity of the experience. Just like any other specific form of anxiety, Maths Anxiety does not remain consistently high for 24 hours a day; it most likely occurs when an individual is using or learning Maths and also increases the closer a person gets to a Maths situation, such as a Maths test. However, there may be some situations that intensify the feeling of anxiety, or even alleviate it slightly.

The amount of anxiety experienced is unlikely to remain at a predetermined level, as many things can influence someone's experience of Maths Anxiety. During some mathematical situations, an individual's level of Maths Anxiety may be strikingly high, being detrimental to them in many ways, whereas within other mathematical situations, they may feel less anxious and more able to cope with the situations they are in. Knowing that Maths Anxiety can be influenced by a variety of factors is useful. If we can identify these, we are more likely to be able to remove, or improve, any detrimental factors and support maths anxious students in the classroom.

So, what can influence Maths Anxiety?

The environment and those who we share our experiences with can have a huge impact on the way we learn, the way we see ourselves, and the perceptions that we create. This can influence many things in relation to Maths, whether this is our attitude towards it, our motivation, or levels of Maths Anxiety. In particular, the role of other people has been linked to the development of Maths Anxiety, and increasing the intensity of the anxiety experienced. Perceptions of other peoples' mathematical capabilities, negatively comparing yourself to others, and interactions with others, have all been found to influence Maths Anxiety. Similarly, timed mathematical situations and specific mathematical content are features of Maths experiences that some have argued can contribute to feeling maths anxious.

This chapter will explore factors that can impact the experience of Maths Anxiety in relation to a few key areas: parents, peers, siblings, teachers, timed situations, and learning specific mathematical concepts. These areas, relating to the role of other people and mathematical situations, are widely researched as to how they can influence Maths Anxiety, usually in a detrimental way.

Parents

The "nature or nurture" debate is complex. In Chapter 2, we looked at whether genetics, our nature, can determine whether someone is more likely to be maths anxious than others. Genetics have been found to account for some of the likelihood of an individual developing Maths Anxiety, meaning that if someone's biological parents are maths anxious, this may predispose them to the experience through their inherited DNA. Not all of the differences found between maths anxious individuals and those who are not maths anxious could be explained by genetics, though. While genetics may impact our cognition, personality, and behavioural traits, it is suggested that our interaction with our surroundings, and those who we interact with, can also influence the way we think, feel, and act - how we are nurtured.

This is what makes the role of parents in relation to Maths Anxiety so interesting, as they are linked to both our nature and nurture. We inherit our genes from our parents, but we are also nurtured by them and are influenced by their language, behaviours, attitudes, and beliefs. We are not able to influence our genetic disposition, but we can be aware of it, and how it may make us more susceptible to certain things, such as Maths Anxiety. How we are nurtured, however, is much less set in stone, so exploring the impact of this upon the development of Maths Anxiety can inform how it may be possible to change this to encourage more positive relationships with Maths.

Our parents undoubtedly influence us, whether this is our attitudes, emotions, beliefs, or behaviour. We are taught so many things by our parents, and others who heavily influence our upbringing, which can leave a lasting impression on us. From how to brush our teeth to which football team to support, the adults who care for us as we grow up have a huge impact on how we view and interact with the world. Watching how they interact with others and navigate through life leaves us with a blueprint to follow, or perhaps a manual of "what not to do."

Either way, we can learn from how our parents interact and respond to various situations, whether this is with confidence and enjoyment, or apprehension and fear. If parents approach a situation confidently and with excitement, such as swimming in the pool, and talking about how enjoyable swimming is, this may be learnt by their children who are witnessing this positive reaction to a situation. This will make it more likely for their child to approach swimming with confidence and excitement, just like their parents.

Sometimes, parents may instead model anxiety to their children through their reactions to different situations. Responding to a situation with concern can prove useful, as it can teach children to stay away from dangerous circumstances, such as an open fire, without them having to experience this danger themselves. This nervous response is mimicked by children, solidifying their understanding of how to react when presented with potential danger.

When parents approach other situations with anxiety and distress that are not life-threatening, however, such as feeling claustrophobic in a small space or hearing a balloon pop unexpectedly, this can also be imitated by their child and can become their learned response when presented with this situation themselves. While the parent may react this way due to their own previous experiences or beliefs, if a child repeatedly sees their parent react negatively towards something, it is likely that this behaviour and associated beliefs are learnt by their child. This then becomes their immediate reaction to the event, with little thought as to why.

> Bartley and Ingram (2017) explained how parents model their own beliefs and attitudes towards Maths in various ways in the home, which can be positive or negative.

The same is thought to be true for behaviours, perceptions, and attitudes towards Maths. Some parents may recall fond memories of learning Maths at school and really enjoyed their lessons. They may have continued to learn Maths as a subject past the age of requirement, such as A-Levels or University. They may remember that learning Maths was quite easy, or if it was difficult, they enjoyed working hard and were motivated to overcome any challenges. If parents share these past experiences with their children through positive discussion of the subject and modelling a positive relationship with Maths, this is likely to have a beneficial impact on their children's behaviours, perceptions, and beliefs relating to Maths.

However, if a parent is negative towards Maths in their behaviour and language, this may be learnt by their child. A parent may have had negative experiences with Maths previously, such as finding it difficult and demotivating, being uninspired, or they may not have had good relationships with their Maths teachers. Any one of these can lead parents to dislike Maths and have negative perceptions of the subject, with the likelihood being that they did not continue learning Maths past the age of requirement. This negative attitude towards, and perception of, Maths can be echoed by their children, who have learnt this from watching their parent's language and behaviour. Negatively discussing Maths as a subject or in everyday life, avoiding Maths completely, or becoming stressed when having to use it, can all demonstrate a poor relationship with Maths that can be embodied by their children.

> Soni and Kumari (2015) conducted research in India with 595 students between the ages of 10 and 15 years, along with their parents (one parent measured per student). They found that parents' Maths Anxiety significantly predicted their child's Maths Anxiety, and their attitudes towards Maths. Similarly, parents' attitudes towards Maths significantly predicted children's Maths attitudes and Math Anxiety.

Research has shown that the stronger the Maths Anxiety experienced by a parent is, the more likely it is that their child experiences this too. Alongside genetics, it has been suggested that this is due to the interactions between a parent and child relating to the subject, or the child observing how parents interact with Maths. Research has also indicated that the more a child is exposed to their parent's Maths Anxiety, the more likely they are to have high levels of Maths Anxiety themselves. This supports the idea that how a parent interacts with Maths can influence how their children will.

> Becker et al.'s (2022) findings indicate that parents' levels of Maths Anxiety can predict their children's Maths performance, even when they are four years old. Higher parental Maths Anxiety predicted lower Maths performance in the Spring term, even after controlling for children's Maths performance in the Autumn term.

It is important to note that not all research indicates a relationship between a parent's level of Maths Anxiety and their child's, so it cannot be an assumption of ours that if a parent is maths anxious, their child will be too. Some studies involving younger children have found no evidence of this relationship, nor a predictive pattern that could suggest a developing correlation. This may be due to the children in these studies being young though, and therefore having less exposure to their parent's Maths Anxiety.

> Jameson (2014) investigated whether a parent's Maths Anxiety was predictive of their child developing Maths Anxiety. In a sample of 91 pupils in the US, aged between seven and eight years old, and 81 of their parents, results showed that a parent's level of Maths Anxiety did not significantly predict their child's.

A further consideration is that parental pressure can lead to Maths Anxiety developing, often by parents who are not maths anxious. This would suggest that although parents are influential in this manner, it is not a parent's Maths Anxiety that leads to their child experiencing their own Maths Anxiety. Instead, parents who are keen for their child to have high attainment in Maths and continue to learn the subject past the age of requirement often place high expectations on their child to achieve in the subject. This may be in the form of additional work to complete at home, or extra lessons given by a tutor. Parents may celebrate the outcome of Maths assessments, rather than the process of learning and the effort that went into the achievement, or be disappointed when their child did not receive high attainment scores. We should also be mindful that high expectations and verbalisations of the importance of mathematical success can place high pressure on students, potentially leading to the development of Maths Anxiety.

Sally's Experience

Sally was quite open when talking about how she felt pressured by her parents to succeed. She felt very aware that her parents had paid for her to attend an independent school, and spoke about how they would regularly ask her about her success in Maths, whether in relation to lessons or tests. Sally felt like she was letting them down, and that these thoughts of pressure and failure would always creep into her mind each day.

> Sari and Hunt (2020) studied 186 child-parent dyads in Turkey. They found that children's attitudes and feelings towards Maths were not related to their parents'. However, parents' attitudes and feelings towards Maths were a significant predictor of children's Maths achievement, but only in Grade 4, not in Grade 3.

It is possible that parental pressure can occur even if the parent is maths anxious themselves. For example, if a parent is maths anxious and has a difficult relationship with Maths, they may place pressure on their child to engage with Maths and succeed, in a bid to ensure that their child does not experience Maths Anxiety like them. At times we hear this being described by parents, not so much in relation to placing pressure on their children, but usually in the context of greater encouragement and support. One of the challenges here is the fine line that can appear between a parent's will for their child to succeed and a sense of pressure that a child may then experience.

> Szucs and Toffalini (2023) suggested that the high levels of Maths Anxiety in students in China may potentially be related to the high amount of parental pressure for them to succeed academically in Maths assessments.

These are interesting ideas to consider, particularly in relation to the nature or nurture debate, but the majority of research does indicate a correlation between a parent's level of Maths Anxiety and their child's. As it is likely that parents can unknowingly model Maths Anxiety and negativity towards Maths to their child, it is not only important to identify how this happens, but also how to best support parents to prevent this from happening in the future.

As teachers, we know the frustration of parents coming into meetings and saying:

> "Maths is pointless, you can use your phone."
> "They're never going to need this outside of school."
> "They won't use Maths once they're 16."

It's even worse when this relates to their own, or their child's, mathematical ability:

> "I was never good at Maths, they're just like me."
> "They won't pass the test, they can't do it."
> "They don't have a Maths brain."

This frustration is heightened when their child is sitting right next to them as they say this. This language places a ceiling on a child's belief that they can achieve, before they've even begun. It is often reflective of having a fixed mindset, with the belief that no matter how much effort is placed into learning Maths, it is pointless because you are born with a certain Maths ability, or lack of, and nothing will change that.

> Fiore (1999) conducted an in-depth study of two maths anxious students and suggested that the damage to self-esteem and performance caused by negative self-talk from key figures, such as parents or teachers, should not be underestimated, irrespective of how long ago it took place.

This is not to say that parents purposely use language that leads to their child developing a negative relationship with Maths. Fixed mindset language can be instilled in parents from when they were young, and it is their genuine belief based on their own experiences. In fact, certain statements such as those above might even be made as a way of parents acting out of empathy, especially if they perceive their child as struggling with Maths. No harm may be meant by these words, or negativity, but language is powerful. It has the ability to strengthen a student's relationship with Maths, or destroy it entirely.

As well as language, a parent's behaviour towards Maths can also affect how their child reacts to mathematical situations. Children may learn to respond negatively towards Maths if a parent models this through their own behaviour, including appearing anxious or stressed.

> Berkowitz, Gibson, and Levine (2021) suggested that maths anxious parents are less likely to engage in "quality number talk" with their children, which can impact their children's development of Maths understanding.

Regular, everyday situations involving Maths, such as finding the right amount of coins to pay for a parking ticket, can cause parents stress, whether this is due to feeling anxious about the Maths involved, or whether they thought they had more coins in their pocket than they actually did. Reacting to a mathematical situation such as this with evident stress or worry, regardless of the reason, can reinforce to their children that this response is normal or expected.

The same is true for situations involving mathematical learning. If parents, maths anxious or not, behave negatively towards the subject, perhaps by avoiding conversations involving Maths lessons, opting to never help with Maths homework, or always using a calculator for the simplest of calculations instead of using mental arithmetic, this can be mimicked by their children. This discomfort can become a learned behaviour, which can affect children having positive relationships with Maths, and potentially giving rise to Maths Anxiety.

> Foley et al. (2017) suggested that maths anxious parents might model non-verbal behaviours, such as facial expressions, gestures, tone of voice, and body language, which may convey negative affect about Maths.

While a parent's language and behaviour can impact how their child responds to Maths, including a response of anxiety, a parent's mathematical success has also been shown to be influential. If a child identifies that their parent has a positive relationship with Maths, one which is celebrated and leads to successful outcomes, it is likely that their child will also develop a positive view of Maths and work hard to succeed in the subject.

For example, a parent may share stories of how they worked hard in school and achieved good results in Maths due to this. They may explain to their child how this has benefited them in their career, as obtaining high results in their exams allowed them to study their chosen subject at University, and their subsequent degree allowed them to be qualified enough to train in their chosen occupation. Similarly, they may model to their child how they use Maths successfully in their everyday life, through managing their money each month with a good awareness of their salary and outgoings, and a calm response to any unexpected bills that come their way.

But what if a parent does not deem themselves to be mathematically successful, or has not gained academic qualifications in Maths? Does this have a negative impact on their child's relationship with, and perception of, Maths? Some research indicates our attainment is related to what we inherit from our parents. There might be some truth to this when it comes to Maths, but it is evident that how parents interact with Maths and how they model this to their children is important.

A parent may share with their child that they were not mathematically successful at school, such as not achieving a pass grade in their GCSE Maths exam, perhaps discussing how challenging it was, or how unmotivated they were. They may also demonstrate finding it difficult to mentally calculate the cost of an item that is marked down by 20% from the original price, perhaps leaning on the use of technology or other people to help them. Their reaction may model distress, concern, or panic. This may make it more likely for their child to mirror this in relation to Maths, and become anxious when using or learning Maths, or expect negative outcomes and therefore not try due to their anticipation of failure.

Research has also found that a parent's job can impact the likelihood of their child experiencing Maths Anxiety. If a parent has a job that involves Maths quite heavily, therefore illustrating mathematical success, their child is less likely to be maths anxious than those whose job does not involve an excessive use of Maths. Even the sector that a parent works in has been correlated to their child's Maths Anxiety, as those with a job relating to teaching, the sciences, health, or business, had children who were less likely to have high levels of Maths Anxiety. In fact, these children were more likely to have high mathematical attainment as well. This again may relate to our genetics, as to work in these sectors, parents need the capability to gain high educational qualifications and have a successful relationship with Maths. As we have discussed in Chapter 4, students with high attainment are overall less likely to be maths anxious. It seems there is the potential for some complex interactions between children's and parents' anxiety and attainment in Maths that are worth exploring further.

It could be argued that parents working in certain job sectors are modelling a successful relationship with Maths and positive outcomes in relation to this, demonstrating Maths utility after compulsory schooling. This can have a positive effect on their children's perception of Maths, making it less likely that they will experience Maths Anxiety. Children with parents who do not work in these sectors were found to be more likely to experience Maths Anxiety and have lower mathematical attainment. This was particularly the case if a parent's job was classified as "skilled" or did not fit into a predefined category. This may relate to the role of socio-economic status in the likelihood of developing Maths Anxiety, which was explored in Chapter 2. Research has shown that students from a low socio-economic background were more likely to experience Maths Anxiety than those from a high socio-economic background. A parent's education level and occupation can impact their family's socio-economic status, which provides further evidence for how a parent's job has the potential to impact their child developing Maths Anxiety.

> Lane (2017) conducted research across 23 secondary schools in Ireland, with 356 participants aged between 15 and 18 years old. The data showed that participants with parents who worked in an occupation that involved the use of Maths had the lowest levels of Maths Anxiety. Comparatively, those with parents whose professions were classed as "skilled" or "other" instead had the highest reported levels of Maths Anxiety.

Researchers have shown trends between parental occupation and Maths Anxiety, but it is important to remember that these are based on samples of people, not the entire population. The relationships observed between a parent's own levels of Maths Anxiety and their child's, or the impact that their educational level or occupation can have upon their child's experience of Maths Anxiety, are unlikely to happen in every instance. There will be examples of relationships where the opposite is true, and a maths anxious parent doesn't have a child who is maths anxious. This does not mean that the research is wrong; it just means that it is not always applicable to each individual situation. After all, we are all unique, and so are experiences of Maths Anxiety.

Rosie's Experience

Rosie's parents worked as a Maths teacher in secondary school and as an architect. Both of these jobs require a good understanding of Maths, and Rosie shared that she felt her parents were very good at Maths, and always found it easy. In contrast, Rosie was maths anxious and said that she would not have a job that relates to Maths in *"any shape or form."* This highlights how a child can be maths anxious, even if their parents aren't.

As these findings are not steadfast rules, it's extremely important not to have preconceptions about who will or will not be maths anxious. However, they do allow us to spot useful patterns and illustrate that parents are likely to be influential in their children's attitudes, perceptions, and beliefs about Maths, potentially leading to Maths Anxiety. Whether this influence derives from our genetic nature, or from how we are nurtured, it is important that teachers are aware of the powerful influence that parents can have on how their children view Maths, and seek to know more about how this can lead to negative outcomes, such as Maths Anxiety, in a potentially intergenerational pattern.

One of the key links between teachers, parents, and students is homework. From a Maths education standpoint, there are various arguments surrounding the purpose and effectiveness of Maths homework. Viewing the publicly available Maths homework policies across schools within a local authority shows just how much variation there is. These can range from a policy that states it is the school's philosophy to minimise formal Maths learning outside of school, to policies that encourage parent-student interactions, sometimes prescribing an exact number of homework interactions per week. It is no wonder that many parents feel uncertain about what they "should" be doing to support their child. Is it best to have regular interactions and to thoroughly read up on the latest Maths calculation strategies being taught in schools, or is there something to be said about distancing yourself as a parent?

> Daucourt et al. (2021) carried out a meta-analysis that showed that parents' involvement in Maths at home is positively related to children's Maths achievement. These findings, however, do not consider the role of Maths Anxiety.

Perhaps it's not so much the intentions, but rather the emotions and behaviours involved in Maths homework interactions that is important in the context of Maths Anxiety. When schools encourage or even require students to engage in Maths homework, this can begin from quite a young age. It is also one of the first experiences that a parent might encounter in terms of supporting (or even just being aware of) formal homework, and many parents will profess to at least one challenging Maths homework encounter. We know from the wider academic literature that parents who exhibit controlling homework behaviours (such as love withdrawal and using commands to pressure outcomes) are more likely to have children with low academic outcomes.

We also know that parental overcontrol, overprotection, conflict, rejection, and criticism can affect childhood anxiety more broadly. It is possible that some parents engage in these controlling behaviours while helping their children with Maths homework, leading to adverse effects regarding their children's Maths Anxiety and Maths achievement. However, a question remains over the extent to which parent-child Maths homework interactions have a causal role to play in the development of Maths Anxiety. There is some evidence for this though. Knowledge of such research findings might support schools and teachers when it comes to advising parents, especially considering that parents will often proactively seek advice.

> DiStefano et al. (2023) observed 40 parent-child interactions on a simulated Maths homework-helping task and measured a range of components to assess the quality of the interaction. They found that more quality interactions were associated with better task performance, but it was only associated with the child's Maths Anxiety, not the parent's. In other words, the overall quality of the interactions was more negative when children were higher in Maths Anxiety.

Studies such as the ones noted here indicate that some interesting things might occur during Maths homework-helping interactions with parents. However, such studies usually suffer from the same limitation as many other studies into Maths Anxiety: they are cross-sectional and do not allow for observation of changes over time. This presents a problem when it comes to drawing conclusions about the way in which key variables might influence (and potentially cause) others. For instance, does it make a difference whether a Maths homework-helping interaction between a parent and child starts off with one or both parties feeling frustrated or emotional, or is it the quality of the interaction and the emotions experienced by the end of the interaction that has the greatest impact on performance and Maths Anxiety? Either way, having an awareness of the emotional state of the parent and child involved in Maths homework interactions is likely to be important to ensure that the interaction starts and remains positive.

> Maloney et al. (2015) found that children learned less Maths when parents with higher Maths Anxiety frequently helped with Maths homework. The level of Maths Anxiety of these children also increased over the course of a school year.

It is likely that many factors come into play when considering the complex interplay between children's and parents' emotions, attitudes, and attainment in relation to Maths, especially once various Maths behaviours are taken into account. While some researchers have looked at the specific role of "number talk," this in itself says nothing about the actual quality of parent-child Maths interactions. Likewise, frequency of interactions based on Maths activities might not be as important as the quality of the interactions themselves.

> Silver, Elliott and Libertus (2021) measured parents' beliefs about the importance of their pre-school children reaching certain benchmarks in Maths. More positive beliefs were associated with children's higher Maths abilities, whereas parents' Maths Anxiety was not a significant predictor. However, further analysis showed that the interaction effect of parents' Maths beliefs and Maths Anxiety was important. For parents with higher ratings of the importance of Maths, their Maths Anxiety was associated with their child's Maths ability. In other words, children whose Maths anxious parents held strong beliefs about the importance of Maths performed best. The frequency of number talk and Maths activities did not affect this relationship.

Studies such as the one above emphasise just how many variables might need to be considered when investigating the relevance of parents in the context of children's Maths outcomes. In reality, it is extremely challenging to systematically research parent-child Maths interactions. As with much research of this nature, there is always a trade-off between experimental control and the validity of the situations in which parents and their children are tested. Often, greater realism results in lower experimental control, whereas a more controlled experimental situation (laboratory-based, for example) can raise doubts whether interactions reflect what they are actually like at home. We also need to be mindful of the age groups being studied and whether measurements and observations are made over time to account for potential changes. Instead, many parent-child Maths studies that look at Maths Anxiety and parent-child interactions are cross-sectional, or involve only a short period between time points.

Peers

Another influence upon how we perceive and engage with the world around us, including Maths, comes from our peers. This may be our friends, classmates, or simply acquaintances. Our peers' actions and beliefs may be shaped by the social norms that surround us; the often unwritten rules followed by the group we are in, which can be affected by culture, values, and beliefs. These are the agreed-upon ways everyone should act and work together, such as waiting in a queue to pay at a shop, or sitting in your assigned seat in a classroom. Observing how our peers engage with social norms in different situations can influence how we then react to various scenarios, how we perceive these, and how we react to these social expectations on our own.

The opinion of our peers is important and influential to us. This includes their opinion of Maths, including important attitudes, such as perceived importance or usefulness, level of enjoyment, or even whether they consider only certain types of people to be good or bad at Maths. While there may be options as to how we solve a question, such as using the grid method to multiply rather than using column multiplication, there is no subjectivity related to the answer – you either did it right, and got the correct answer, or you did it wrong, and are incorrect. It is perhaps this that leads Maths to be seen by some as socially divisive – either loving it or hating it. While some people display no particular attitudes towards Maths, the majority of people, including young and adult students alike, seem to have a strong opinion:

> "What do you think of Maths?"
> "It's my favourite lesson ever!"
> "Not a fan."
> "I love it!"
> "Not for me."
> "It's fun!"
> "Worst thing ever."

Our peers' opinions and perceptions of Maths are often shared with us, and are influential to us, through their use of language and behaviour. We may adopt and mimic this behaviour and language as we want to "fit in" with a group. It is also possible that students' views on Maths are affected by peer pressure; that they feel somewhat forced into holding a particular attitude towards Maths. This might be an outwardly facing attitude for the purpose of following the crowd, or wanting to avoid potential consequences of deviating from the norm. Over time, an outwardly facing attitude might become more than that, gradually changing a student's internal belief system to the point where they fully take on board the views of the wider group.

While empirical evidence of this in the context of Maths is scarce, the field of social psychology has a plethora of studies demonstrating the importance of social influence, particularly in relation to conformity and normative social influence. Whether you are a teacher or not, we have all been involved in, or heard conversations about, others' views on Maths. We are also regularly exposed to representations of Maths within the media; for instance, the stereotype of the "Maths geek" is commonplace in films and television. In the case of schooling, students are surrounded by peers on a daily basis, almost always the same age and often of a similar socio-economic status. Thus, the relevance of social influence in the context of Maths and the potential to shape an individual student's attitudes towards the subject should not be underestimated.

> Gest and Rodkin (2011) identified how children can mimic their peers' language, behaviours, and beliefs to reflect social norms.

Let's look first at how Maths Anxiety may be influenced if an individual's peers have a negative relationship with Maths.

As a regular occurrence within the classroom, classmates may talk about how Maths is too difficult, boring, or pointless. They may talk negatively about their Maths teacher, or about how every other subject is much more interesting and useful in everyday life compared to Maths. Over lunch, a student may share that they had a horrible Maths lesson earlier in the day, or on the way to school they may make a comment about dreading the Maths lesson that they have later on that day. Often, these conversations consist of throw-away comments without much thought behind the effect that they can have on someone. Students making these comments might have a negative relationship with Maths and simply don't enjoy engaging with the subject. They may be experiencing Maths Anxiety themselves, or they may be using defensive pessimism as a coping mechanism in a stressful situation, where an individual uses negative language to prepare themselves for the potential of any unfavourable outcomes. Another reason behind these comments may be that they believe it's what their peers want to hear, and they are saying it due to experiencing peer pressure.

> Hatfield, Cacioppo, and Rapson (1992) explored the notion of emotional contagion, referring to when an individual's feelings can transfer onto those around them. Individuals may mimic others' language, expressions and behaviour, which reflect their emotions.

Regardless of the reason behind them, these comments can have a powerful effect on their audience. Utilising this language on a regular basis can create a shared negative perception of Maths among peers, which can become an embedded, mutual belief towards the subject. This co-rumination, where peers explicitly discuss their negative emotions towards Maths, can encourage an individual to do the same, particularly if their peer is a close friend. If a maths anxious individual is surrounded by this language from their peers, it is likely to enhance their feelings of anxiety and have the potential to justify their fearful response to Maths. This language may also be influential in Maths Anxiety developing in the first place, which may strengthen in a cyclical pattern the more they are exposed to their peer's negative language surrounding Maths.

This emotional contagion has been found to influence an individual's intensity of Maths Anxiety, as it provides an opportunity to openly discuss a shared experience through the understanding of having mutual beliefs, attitudes, and perceptions. If a maths anxious individual aligns with peers who are also experiencing Maths Anxiety, this provides them with an opportunity to discuss this, normalise it, and share experiences. As such, this can intensify the anxiety felt, as they are being given validation from their peers.

> **Ali's Experience**
>
> Ali seemed enthused when speaking about her friendship group in her class, as they *"didn't like Maths either."* She shared that she and her friends didn't really talk about Maths often, but if someone felt that something was difficult, all the group supported them and shared that they found it difficult to. Whether they did this to make the individual feel better or not is unclear, but this does indicate the role that peers can have upon emotions towards Maths.

It is also feasible to consider that maths anxious students group together, arguably creating a shared identity. In fact, listening to a group of people having a conversation about Maths sometimes reveals interesting things. After quickly gauging others in the group, a single person might speak out about feeling maths anxious, or being poor at Maths (or both). This can often escalate into the others within the group with claims that they are the same. This shared identity and willingness to open up in an environment that is deemed psychologically safe appears to act as a safety net for some people. Such a scenario is commonplace when groups of students meet for the first time, especially adult students who are embarking on a programme of study that includes a Maths component, but one that does not necessarily have Maths at its core. Typical examples include not only certain apprenticeships but also social science degree programmes that involve the study of statistics, and nursing and healthcare programmes. This is one of the reasons why tackling these issues head on at the start of a programme can be so important.

> Kim, Shin, and Park (2023) researched whether a maths anxious peer can influence the Maths Anxiety experienced by an individual. Through longitudinal research over the course of an academic term, they found that an individual's Maths Anxiety became more similar to that of their peers. This suggests that a maths anxious peer can be influential to an individual's own experience. Interestingly, they did note that this occurred within already formed friendships among peers, as they did not form new friendships on the basis of sharing the experience of Maths Anxiety.

Negativity towards Maths is likely to be mirrored in our peers' behaviour as well. They may avoid the subject altogether, or have a lack of motivation to engage with it, leading to a lack of effort when using or learning Maths. This may be reflective of their own experience of Maths Anxiety, or a simple dislike of Maths. In the classroom, a student's peers may choose to do the minimum amount of work required to complete a mathematical task, or not complete the task at all. They might even engage in a completely different task simply to avoid the task at hand. Such behaviours can be influential to an individual observing them, as they can be learnt and mimicked, perhaps due to peer pressure or an individual's desire to be like their peers. Of course, it can also be the case that students spot tried and tested ways that their peers adopt to temporarily avoid facing Maths as a coping strategy, subsequently adopting the same behaviour themselves.

It should also be considered that an individual may observe this negative behaviour from their peers and subsequently identify that it is detrimental to their peer's learning and future choices, as it may have led to poor Maths grades, for example. In this way, a peer's negative relationship with Maths could positively influence the student, as they are then motivated to avoid these outcomes and work hard to ensure that this does not happen to them. However, the extent to which this scenario plays out in the case of the maths anxious observer is questionable.

> Garba et al. (2020) found that the negative behaviour and language of peers, such as being disruptive in lessons, had a direct, "intensifying" effect on Maths Anxiety. Students who dominated the lesson in terms of engagement also intensified Maths Anxiety in others.

While the inclination might be to think of peer influence as a negative force, it can be really beneficial in building relationships with Maths. It can guide us through the network of social norms and prompt us to adopt behaviour and language that have positive consequences and successful outcomes, such as working hard in lessons to achieve the career path we dream of. Peers may find the subject quite easy to understand and use in a variety of different situations, or they may find it quite interesting. They may share these thoughts regularly in conversation with one another, stating that Maths is the best subject out of the whole curriculum, and it's the most important one to learn, before listing off multiple reasons. In the classroom, peers may talk about how easy and straightforward the lesson of algebraic expressions was, and how the teacher explained it in such a relatable way.

Outside the classroom they may also use language to express their positivity towards Maths, such as how useful it is to help them work out the 25% discount at their favourite clothes store, or how good it feels to get a calculation in their homework right after what seems like hours of trial and error. Being part of these conversations, or overhearing them, on a regular basis can lead to an individual developing these positive attitudes towards Maths as well.

The same can be true for our peers' positive behaviour, as they may model high levels of motivation and engagement with Maths through the actions that they take. For example, a successful relationship with Maths may be shown through peers always putting their hands up to answer questions during Maths lessons, or persevering when the work is challenging, resulting in a sense of accomplishment. These behaviours provide students with positive examples to follow, and if they witness successful outcomes as a result of these, such as achieving good test scores, or receiving positive comments from teachers or parents based on exerted effort, this reinforces the value of having a positive attitude towards Maths. If the majority of a student's peers behave this way, they may mimic this social norm and seek to reflect this in their own behaviour.

> Garba et al. (2020) also found that the positive behaviour and language of peers, such as modelling success and achievement, or providing positive advice, can reduce feelings of Maths Anxiety.

But, what happens when peers' positive behaviour and language doesn't align with a student's true beliefs?

If an individual finds Maths challenging and has negative perceptions of it, but the general consensus of their peers is that Maths is enjoyable and easy, this can cause internal conflict, and potentially, Maths Anxiety. For example, if a group of students were working on a task in a Maths lesson, and they were discussing how fun or how easy it was, a maths anxious student may feel isolated as they are instead finding the task challenging and feel unable to share their thoughts and emotions with their peers. This may lead to feelings of pressure on a maths anxious individual to keep up with the group's pace, despite not understanding the concept. It may also make them feel uncomfortable to ask questions when needed, which might exacerbate their experience of Maths Anxiety and their lack of self-belief in their mathematical capabilities.

Sean's Experience

During a Maths lesson, Sean appeared very quiet and reserved and didn't speak in the group that he was working in. Afterwards, when asked why he didn't participate in any discussion, he shared that he felt overwhelmed. He felt that everyone else in the group found the task really easy, but he found it challenging. Sean felt different to his group, and that he couldn't contribute as he was *"not as good as them."*

Therefore, while peers' positive language and behaviour towards Maths can be beneficial in positively influencing those around them, it can also have the opposite effect and lead to feelings of uncertainty or panic, or feeling completely overwhelmed. Again, this may influence the development of Maths Anxiety, potentially in a cyclical way; the more an individual feels isolated and different from their peers' successful relationship with Maths, the more this feeds into their anxiety, which in turn continues to widen the perceived gap between themselves and their peers.

The conflict between a maths anxious individual's perception of their peers' success and relationship with Maths to their own can also lead to comparison, with detrimental effects. A maths anxious individual may be likely to compare themselves to their peers in a negative way, which may relate to attainment, motivation, or enjoyment of Maths, as a few examples. If others are perceived by a maths anxious individual to be engaging positively with Maths, achieving high marks on tests, and receiving praise from their teachers, this may highlight to them how their own attainment and relationship with Maths is not as successful. This comparison may serve to increase a maths anxious individual's lack of self-belief, lack of motivation and negative attitudes towards the subject, as well as their feelings of panic and anxiety.

> OECD (2015) reported on PISA 2012 data, showing that students tend to feel more anxious about Maths when most of their peers perform better than they do.

As peers are usually an inherent part of a student's everyday life, this comparison can be hard to avoid. Maths anxious individuals may negatively comment on these comparisons, through language such as:

> "They're way better at Maths than me."
> "How do they get it so quickly?"
> "What's the point, I'm never going to be like them."

The use of this comparative language may indicate that the individual is finding it difficult to adopt the positive language and behaviour of their peers, that they feel inferior to them, and that they are perhaps experiencing Maths Anxiety. Being mindful of this type of language in the classroom and at home may allow teachers and parents to support these individuals to improve their perceptions of themselves, regulate their perceptions of their peers, and improve their self-belief and self-talk. This in turn may improve an individual's experience of Maths Anxiety and support them in developing a more positive relationship with Maths.

It is clear that the role of peers within Maths Anxiety is complex, as both positive and negative language and behaviour can influence how an individual perceives Maths, and also themselves, and others, in relation to Maths. Knowing this, however, and not just working with the assumption that only negative behaviour and language can have a detrimental impact on maths anxious individuals, will work towards a better understanding of how peers can influence the individual experience of Maths Anxiety. Taking a more positive stance, we will come back to the idea of peer support in Chapter 6, as the way in which young people have been found to perceive social support from peers, compared to parents and teachers, provides food for thought in the context of Maths Anxiety.

> Bokhorst, Sunter, and Westenberg (2010) reported survey results of 655 young people in The Netherlands. Similar levels of social support from classmates and teachers were perceived among 9-15 year-olds. However, in 16-18 year-olds, whilst classmates were perceived as equally supportive as parents, teachers were perceived as less supportive.

Siblings

Like our peers, siblings can also have an impact upon our relationship with Maths, and the experience of Maths Anxiety. While there isn't a vast body of research into how siblings can influence Maths Anxiety, it is generally agreed that siblings can be one of the first role models a child has. This means they can learn from and mimic their sibling in the home environment, before an individual even begins their years at school. With such an early influence on how an individual views and interacts with the world, including Maths, it is important to explore the potential role of siblings within the experience of Maths Anxiety. The influence that our siblings may have on us can be positive, but it also has the potential to have a detrimental effect on our perceptions, beliefs, and behaviours relating to Maths. This can occur when our siblings are modelling either negative behaviour and language or positive, and may be more likely if our siblings are older.

If siblings regularly use negative language to discuss Maths, and their behaviour towards the subject is detrimental, this can be learnt by their siblings, who can then form their own negative perceptions and relationship with Maths. If an individual's brother or sister often discusses their dislike for the subject for example, or shares that it's too difficult, this can become a learnt and shared use of language at home, which becomes the norm. Similarly, if an individual's sibling models negative behaviour, such as refusing to complete their homework, or ignoring their parents when asked for their help to accurately measure ingredients when cooking, this can be imitated.

> **Olivia's Experience**
>
> At home, Olivia felt that nothing positive was ever said about Maths. Her and her two sisters would always complain about homework, relating to how hard it was, or how boring it was. They often spoke about never needing Maths after school, and how pointless it was. Olivia said she *"probably didn't hate Maths as much as them,"* but still didn't love it because it made her feel panicked.

If a sibling's language and behaviour demonstrate that they should respond to Maths with fear, dislike, worry, and anxiety, this can be a learned response. For example, if a child is watching their brother or sister go into a state of panic when having to learn their times tables off by heart or sees that they are anxious on the way to school because of their upcoming Maths lesson that morning, this may be viewed as an accepted reaction to Maths, and imitated by the individual. Seeing this at home can instil these behaviours in an individual and inform their own response to Maths, which can also have an impact upon how they use and learn Maths in the classroom. This may not always be the case though, as if an individual's already established beliefs about Maths don't align with their siblings', this imitation is less likely to occur.

> van der Vleuten, Weesie, and Maas (2020) explored whether siblings are influential upon an individual's choice of which subjects to pursue in their further education. They found that younger siblings were more likely to choose subjects or fields similar to what their older siblings did. This was more likely to occur if they were the same sex, and this was found regardless of various family characteristics.

Alternatively, siblings may instead model a positive relationship with Maths through, as an example, always completing their homework on time and engaging in the process of learning at home, rather than shying away from the task at hand. They may talk about Maths in a positive light, sharing how enjoyable their Maths lesson was earlier that day, or how they are looking forward to their next lesson because the topic they are doing is really interesting. This can have a positive influence on an individual and foster the development of successful perceptions of Maths. If an individual's established beliefs about Maths oppose their sibling's positive view however, such as instead finding Maths difficult, or perceiving the subject to be boring or unenjoyable, this can lead to feelings of nervousness, anxiety, and self-doubt, especially when they listen to, or observe, their sibling's mathematical success. This lack of alignment can accentuate an individual's negative relationship with Maths, and any feelings of anxiety towards using and learning Maths.

> Wang et al. (2014) found that the correlation in Maths Anxiety between identical twins was much stronger than that observed between non-identical twins. They further reported that shared environmental factors between siblings moderately predict individual differences in Maths cognition, but not Maths Anxiety. Instead, a combination of genetics and non-shared environmental factors can predict an individual's Maths Anxiety.

This disparity between beliefs can also lead to comparison between an individual and their siblings. If an individual feels that their sibling, or siblings, are better at Maths than them, or that their siblings have a more successful relationship with it than they do, this can influence their own relationship with Maths, and the experience of Maths Anxiety. This can negatively impact their self-belief, enjoyment of using and learning Maths, and their motivation to succeed, leading to feelings of inadequacy.

The negative effects of sibling comparison may be strengthened further when parents highlight and vocalise any differences between siblings in terms of successfully understanding Maths and being successful when using and learning Maths. For example, if a parent asks a child why they don't like Maths and shows concern that this is "different" to their siblings, because their siblings do like Maths, this can lead to feeling inadequate and associated feelings of failure, and potentially panic and anxiety in relation to Maths.

Malik's Experience

Malik did have motivation to learn Maths, but he was still maths anxious. He often spoke about the importance of learning Maths, because his older brother was doing so well at university and he needed a B in Maths to get in. He rationalised that he needed to work hard as well and do well in Maths, but that this did cause him anxiety because he felt that he wasn't as good at Maths as his brother. Malik also felt that his parents knew this and didn't have the same expectations of him as his brother, which made him feel bad about himself.

This raises several questions about the extent of shared experiences between siblings, especially twins, given that twins provide a useful way of assessing the role of genetics. Are siblings likely to have the same interactions with parents when it comes to Maths homework, for example? Are siblings likely to have the same opportunities to access resources, such as a private Maths tutor, for instance? Do such similarities depend on age differences (which of course are non-existent in twins)?

Discussion regarding the role of siblings also feeds into the broader debate of the role of socio-economic status and Maths attainment. For example, families with a higher socio-economic status tend to have fewer children, which could mean more money and time available to support Maths learning. In contrast, a large family with a low income may have few resources available for additional support. We should also consider the possibility that family dynamics change as more siblings are introduced, including less time for parents to play one-on-one number games for example, or to support individual children with their times tables or Maths homework. More siblings would also mean greater complexity regarding potential interactions between siblings themselves, whether that means helping each other out with Maths homework, or competing for their parents' support with it.

While there is not a wealth of literature that details how siblings can influence an individual's experience of Maths Anxiety, it is important to understand not only how complex this influence can be, but the negative effect it may have on an individual's relationship with Maths. Learning to adopt negative language and behaviour, or feeling misaligned with a sibling's positive and successful relationship with Maths, can be detrimental and must be considered as a potential influence when exploring an individual's experience of Maths Anxiety.

Teachers

Teachers are perhaps one of the most researched influences of Maths Anxiety, as they are consistent role models for students throughout their educational journeys, with the power to impact their emotions towards learning and using Maths, and the success of achieving their goals. The social interactions that students have with their teachers in the Maths classroom can influence their beliefs related to Maths, as well as their perceptions of, and relationship with, the subject. Teachers can play a crucial role in mitigating Maths Anxiety through fostering a supportive classroom culture, based on positive shared beliefs, an atmosphere conducive to learning, and the promotion of communication and collaboration. However, teachers may also influence the development of negative attitudes towards Maths, and Maths Anxiety, due to a classroom culture fuelled by situations involving detrimental language and behaviour, a lack of student agency and dialogue, and teaching styles and choices deemed as unsupportive.

> Ismail et al. (2022) carried out research with 20 high maths anxious students in Nigeria. Using a photo voice methodology alongside semi-structured interviews, students reported several ways in which teachers either contribute to, or minimise, Maths Anxiety. For example, fast-paced Maths lessons, providing too many notes and exercises,

> disparaging comments, feeling overly punished, and being ignored, were all reported to intensify Maths Anxiety. Findings also revealed that teachers' good use of learning facilities, consideration of pace, and motivational statements and support were reported by students to minimise Maths Anxiety.

Many things can influence the classroom culture that a teacher creates in their Maths lessons, including their language. A teacher's language can affect how their students relate to, and learn, Maths, but it can also affect their emotions and perceptions towards it, and impact the experience of Maths Anxiety. If a teacher talks about the subject and its content positively, this is likely to demonstrate a positive mindset towards Maths to their students.

"Algebra is just like a puzzle, it's so fun."

Through sharing these positive comments, a student's perception of a mathematical concept can change from being something so unfamiliar, such as algebra, to something that is relatable, and achievable, like puzzles and patterns.

It is important that teachers don't undersell a concept though.

"Dividing fractions is so easy."

This can make a concept appear really easy and achievable but can heighten feelings of anxiety. When teaching a straightforward concept with an easy procedure to follow, a teacher may be looking to provide comfort and support to students by indicating that everyone can achieve the lesson's objective. This can, however, make a student feel isolated and uncomfortable to ask for help if they don't understand the concept or task, as they may feel that the teacher is anticipating everyone to understand and succeed in the lesson, leading to feelings of heightened anxiety.

On the other hand, if a teacher uses negative phrasing, this can make a mathematical concept or task appear too difficult and challenging.

"This is really hard, especially because you need to remember to convert the units of measure."

If a teacher is telling their class that something is difficult, with little else added, students are most likely to perceive this concept or task negatively, with the belief that they are unlikely to achieve. This can negatively influence a maths anxious individual in particular, as this can be detrimental to their self-belief and motivation, further impacting feelings of panic and overwhelmedness. Of course, difficulty in itself is not the issue, but rather the way in which this is communicated. A narrative around the difficulty of a task can result in quite different approaches to it. We often face difficult tasks, and it is the way in which we perceive the tasks, and ourselves, that ultimately determines how we approach them. An acknowledgement of a Maths task being tricky shows to the class that everyone faces Maths challenges. The message that it's fine, and often normal, to stumble along the way, can reinforce a growth mindset and the belief that struggle can actually support our learning.

The language a teacher uses towards a student's success or failure in the classroom can also negatively influence students' beliefs surrounding Maths. By celebrating and praising the success of a student, this can be motivational to others.

"Excellent, you've got all of the answers correct. Well done!"

It can also increase the likelihood of negative comparison though and make other students feel less capable and negatively influence their self-belief, as they do not feel as successful as their peers. This may also place an emphasis on the outcome of learning, rather than the process of learning, which can again strengthen the feeling of pressure and anxiety.

Highlighting any shortfalls of a student, and their failure, can also have this negative effect.

"You only got 2 right, what happened?"

Comments such as these can unsurprisingly increase negative perceptions of Maths and influence the development of Maths Anxiety. They can make a student feel that they did not do their best, and that Maths is too challenging for them to achieve the expectations of their teacher. Feeling overwhelmed, fearing mistakes, and having an anxiety-fuelled response can also be strengthened by comments referring to someone's lack of understanding or poor mathematical attainment. This can be made worse if these comments are shared in front of others, as a student may feel ashamed of their lack of understanding and again, compare themself detrimentally to others within the learning environment, leading to feelings of anxiety.

> **Elijah's Experience**
>
> Elijah could recall instances from his years of school where he felt that teachers thought he was incapable of doing well in Maths. These ranged from commenting on his mistakes in front of the class, to comparing the amount of work he had completed (and tried hard to complete) to his peers. Whether this sense of judgement was perceived or actual, the impact of this was real upon Elijah's experience of Maths Anxiety.

Negative comments about a student's lack of understanding can also influence their peers, as witnessing this language in the Maths classroom can create a classroom culture of negativity and fear of being shamed in front of others. A student may hear their classmate being told by their teacher that they didn't achieve a good grade on their last test, which can increase their own anxiety and worries that this might happen to them if they do not achieve. This can place pressure on students and create a detrimental shared belief of expectations that are unrealistic.

If a teacher has commented on a peer's lack of success as they achieved a grade of 70%, for example, this is likely to infer that a student needs to achieve higher than that in order to avoid negative comments about their attainment. This is unlikely to be achievable for all students in the classroom, and whether these expectations are explicit or not, they can create feelings of dread and panic and increase the anxiety experienced around failure and preempting failure in Maths.

> O'Toole and Plummer (2004) found that social interactions within the classroom had an influence on the mathematical development of students in Australia.

Alongside their language, how a teacher engages with Maths can be influential to the students they teach. As such, their behaviour has the potential to impact whether students form positive relationships with Maths, or negative ones, including the experience of Maths Anxiety.

If a teacher is self-assured when teaching and using Maths, this can positively influence students as it can demonstrate enjoyment, confidence in understanding and interpreting mathematical content, and a successful relationship with the subject. Demonstrating humility in this regard is important though, as if a teacher can appear arrogant in their approach to Maths, this can be disparaging to students who may not have a similar confidence level, and serve only to distance their students from mathematical success and strengthen the belief that they do not have the ability to achieve.

Similarly, a teacher who masks their own Maths Anxiety might not necessarily do so convincingly. This has the potential to prolong a teacher's level of stress, without dealing with their inner feelings about Maths, but we also have to be open to the possibility that students are quite good at recognising when teachers lack Maths confidence. We will touch on this in more detail in Chapter 7.

How a teacher responds to making a mistake when solving a mathematical question is also important to consider, as their reaction to this is modelled to their students. Rather than acting upset or annoyed with themselves, a teacher accepting their mistakes and identifying positive ways forward can be a really beneficial modelling experience for their students. On the other hand, if their teacher is not responding appropriately to making mistakes in the Maths classroom, it can make it difficult for their students. For example, if a teacher becomes frustrated or annoyed at themselves for making a mistake, and makes excuses for this, or appears timid when doing difficult calculations as they are fearful of making mistakes, this behaviour can have a negative influence on their students. If a student emulates this behaviour, this can become a learnt response and be detrimental to their relationship with Maths. This can also develop feelings of anxiety when mistakes are made, leading students to be fearful of trying to achieve in case they are wrong, and perhaps learning to preempt failure before even beginning to solve a mathematical problem.

A teacher's behaviour and actions can also model the importance of learning and using Maths to their students. By providing real-life examples, taking the time to explore how Maths is useful in a variety of situations, and how to successfully use it in a range of contexts, a teacher can demonstrate the usefulness of Maths and build confidence in their students. However, if a teacher is keen to get the lesson over with, or avoids challenging mathematical discussions, perhaps due to a lack of confidence or a lack of enjoyment of Maths, this demonstrates a lack of appreciation of the importance and usefulness of Maths. This may also be shown through a teacher relying on the use of calculators to solve all calculations, rather than using mental or written methods, or regularly using generative AI for answers. These hindering behaviours can be modelled and mimicked by students and perhaps lead to an increase in anxiety when students actually have to engage with Maths without technological support.

Teachers don't aim to model a negative relationship with Maths, but it can happen through the smallest of comments, or the quickest of actions, which are ultimately linked to their own relationship with Maths and their experiences. Whether this is a regular occurrence, or whether this is a one-off due to the teacher having a bad day, this can have a really powerful effect on students, and on Maths Anxiety. Therefore, to truly understand how a teacher can influence the occurrence of Maths Anxiety, and what can be done to ensure this is positive, it's important to identify what can impact a teacher's relationship with Maths, and the influence that this can have upon their students.

Teachers who are not Maths specialists may be more likely to have a negative relationship with Maths compared to teachers who do specialise in Maths. Secondary school Maths teachers teach Maths throughout each and every working day, which is likely to solidify their confidence with Maths and how to teach it effectively. It is also likely that they became a Maths teacher due to having a positive relationship with Maths in the first place, alongside confidence in its applications. Alternatively, a primary school teacher teaches subjects across the curriculum, from Maths to English, to Art to Geography. This means that they must have a good understanding of how to teach a variety of subjects, but that they may not have a specialist understanding of Maths, nor specialist training of how to teach it in the classroom.

While it should not be assumed that a primary school teacher will have a negative relationship with Maths because they do not specialise in the subject, they do spend comparatively less time teaching and engaging with the subject compared to specialist teachers, which may make it more likely. If this is the case, this might suggest that negative relationships with Maths may be more likely to be modelled in primary school than secondary, which can have an early and detrimental effect on a student's schooling.

This may be interlinked with a teacher's mathematical competence and subject knowledge. A Maths specialist teacher in secondary school is more likely to have higher Maths qualifications compared to a primary school teacher. Most teacher training providers in the UK, for example, require a secondary school Maths teacher to have an A-Level in Maths, if not an undergraduate degree in Maths, or a degree which is Maths-related, such as Physics or Engineering. If a provider doesn't request a specific undergraduate degree, they usually require the prospective teacher to demonstrate their subject-specific knowledge in order to gain a place on the training course. A primary school teacher, however, is only required to hold a pass grade or higher in Maths at GCSE level.

Secondary Maths teachers are required to teach to a higher level than primary school teachers, which demonstrates why the requirements are different. For example, a primary school teacher may look at introducing equivalent fractions in Year 3, while a secondary Maths teacher may teach the application of algebraic fractions in Year 13. Ensuring that all Maths teachers in secondary schools have at least an undergraduate degree relating to Maths will likely allow these teachers to feel highly skilled in their knowledge, which may translate to a positive relationship with the subject in the classroom. However, this is a highly politicised debate, with many questions around resourcing that we will not go into here.

Good, Rattan, and Dweck (2012) surveyed a large sample of university students in the US. They found that students' sense of belonging (one's feelings of membership and acceptance in the Maths domain) predicted students' intentions to pursue Maths in the future. That is, a lower sense of belonging in Maths was associated with less intention to pursue Maths in the future. To address this, the authors call for Maths learning environments that develop students' sense of belonging and communicate a growth-mindset message, especially to support females.

It can be argued that primary school teachers don't require a higher level of formal Maths qualification as they don't teach mathematical concepts in the same advanced detail. However, the requirements to teach Year 6 Maths lessons in primary schools compared to Year 1, for example, are vastly different. A Year 6 teacher may teach 10 and 11 years olds how to successfully add numbers with up to two decimal places, while a Year 1 teacher may teach five- and six-year-old children how to add either one- or two-digit numbers up to the value of 20. While the job specification is the same, the knowledge requirement differs. This can cause a strained relationship with Maths if a teacher experiences difficulty with understanding or applying certain mathematical concepts, yet they are required to teach them.

If a teacher feels uncertain about their own mathematical capabilities, this can impact the way they present mathematical concepts to their students, and they may model a lack of confidence in how to respond to Maths. This can impact their students' perceptions of Maths, and potentially their Maths Anxiety as well. If a teacher regularly makes mistakes in front of the class and doesn't have confidence in their own understanding, this can influence the development of Maths Anxiety in their students and create a lack of belief in the teacher. If students don't believe a teacher is capable of helping them learn Maths, this can strengthen the perception that Maths is too difficult to use successfully, and enhance feelings of worry, panic, and anxiety when faced with a mathematical situation themselves. We expand on this in Chapter 7.

A teacher's negative relationship with Maths may also relate to their experience as a teacher. If a teacher is new to teaching Maths, they are more likely to feel less confident in delivering their lessons compared to teachers who have taught for many years. While this is a normal process to go through, this can affect a teacher's relationship with Maths, leading to a lack of confidence and uncertainty. A newly qualified teacher or an early career teacher may also feel a lot of pressure in their teaching performance as well, as they are observed regularly and have additional targets to meet. This pressure may be transferred into the classroom, negatively impacting their own relationship with Maths, but also their students'. This may not be related to a teacher's subject knowledge and understanding of Maths at all, as it may instead be reflective of learning the skills required to be a teacher, such as behaviour management or having a good pace of learning within a lesson. Nevertheless, this uncertainty can be modelled through a teacher's behaviour, such as showing hesitation and a lack of self-belief for example, which can be learnt by the students they teach and impact how they view, and respond to, Maths.

> Harper and Daane (1998) found that, among several factors, an emphasis on right answers was associated with causing Maths Anxiety.

A further way in which teachers can influence how students respond to learning and using Maths is based on the decisions they make as to how they teach the subject. Specific pedagogical choices, referring to how a teacher teaches a concept, may negatively influence students' relationships with Maths and influence the experience of Maths Anxiety. If a teacher's lessons focus on the importance of rote memorisation and learning procedural knowledge, rather than focusing on a student's development of conceptual understanding, this can lead to negative associations and anxiety with Maths. In a lesson focusing on how to multiply fractions, for example, a teacher may just teach the procedure of multiplying the numerators, and then multiplying the denominators, with little discussion as to why. While this will give the correct answer, this procedure isn't always as straightforward. For example, if multiplying mixed numbers, what should students do with the whole number? By teaching the concept behind multiplying fractions and what actually happens, whether through visual representations or physical manipulatives, this makes it more likely for students to acquire the knowledge needed to apply their understanding to a range of situations. With only an understanding of the procedure to follow, and rote memorisation of number facts to help them, this may elicit feelings of anxiety and being under pressure, especially if what they've learnt isn't conceptual or applicable to a wider variety of mathematical situations.

> Ashcraft and Krause (2007) suggest that a non-supportive teacher is one of the key contributing factors in the development of Maths Anxiety.

The resources and activities used within lessons can also impact how a student engages with the mathematical content, and their beliefs towards Maths. If there are a variety of resources available, may feel that they can understand a concept using these resources, and that they have the student agency to explore this. For example, if a teacher ensures that there is a variety of manipulatives available in the classroom, such as Base 10, numicon, and Diennes, and models how to use these to help students understand the concept of place value, this encourages active learning and participation. A teacher may also have a variety of visual representations of a 2-digit number, whether this is on a Base 10 chart, through pictorial representations of counters, or by abstract numbers themselves. Again, this creates a classroom culture based on exploration and discussion and provides various opportunities for students to engage with the lesson.

Without an array of supporting resources, understanding the concept can be quite challenging for some students. In lessons where only abstract concepts are used, for example, without any other representations of the concept, and a focus only on the written questions in a textbook to answer, this can be disengaging to students. It can lead to students finding it difficult to understand a concept, feeling limited in their choices and agency, and therefore feeling anxious in this situation as they feel unable to achieve.

> Yanuarto (2016) conducted a qualitative case study, where university students recalled that their Maths Anxiety developed due to restrictive teaching styles and a teacher's imposed authority on them.

In addition to this, having an authoritarian approach, where there is a lack of student agency in lessons, can reduce students' enjoyment, leading them to feel unheard by their teachers, and elicit negative emotions towards Maths. Research has shown that rigid and structured teaching styles, where lessons focus on independent work without any exchanging of ideas, may make a student more anxious compared to when lessons promote collaboration and discussion. Promoting independent work without any discussion can be isolating to students and potentially increase their feelings of anxiety when faced with a mathematical situation or question that they feel they cannot answer successfully.

This is strengthened when a student believes that their teacher is unhelpful when answering their questions. If a teacher continues to answer a question in the same way or explains a concept using the same method or representation, this is unlikely to support a student's understanding, leading them to feel unsupported and unable to achieve. If a student feels that they are not able to understand what, or how, a teacher is explaining a concept to them, this can further increase the likelihood of them developing Maths Anxiety.

> Finlayson (2014) investigated the reported causes of Maths Anxiety among adult students. They found that 17 out of 30 participants linked their Maths Anxiety to a fear of failure, and 28 out of 30 linked it to the teaching style they experienced. This included the pace being too fast, an authoritative style, and not checking students' understanding.

If a teacher focuses too much on the outcome of a lesson, and completing lessons to ensure they have covered the curriculum objectives within a certain amount of time, this can also generate feelings of pressure, concern, and anxiety. Of course, the reality is that there are often external demands placed on teachers, meaning they feel pressure to work through the Maths curriculum at pace. In fact, many teachers will say that this is one of the contributing factors to Maths Anxiety in schools; both in terms of the pressures experienced by teachers but also fewer opportunities to cover mathematical concepts in depth or to go over them again when needed.

> Helmane (2016) conducted research using a Likert-type scale questionnaire with 124 pupils between eight and nine years old. 36% of participants shared that they had negative emotions towards Maths, including anxiety, because they felt that their teacher had an authoritarian style and did not listen to them or allow them to be active learners.

Grouping students can also impact their relationship with Maths and influence Maths Anxiety. Promoting mixed-ability groupings, where a confident student works with a student who finds Maths difficult, can enable fruitful discussion. If students are grouped according to their ability though, where students who find Maths difficult are put in a group together, their access to certain content and discussion in the lesson is limited. This may have a negative impact on their ability to discuss and understand mathematical concepts, but also on how they view themselves if they are aware of the group they are in. This can also cause anxiety in students, as they may identify that they are being categorised as a student who finds Maths difficult, which they perceive to be an obvious label to their peers in the class. This can cause anxiety in relation to feeling judged by their peers, and their peers' perception of their capabilities.

> ### Layla's Experience
>
> During a conversation about her previous experiences in learning Maths, Layla recalled being moved to the "support table" and being removed from the class to work in a group with the teaching assistant. Even though she was young at the time (she guessed around six or seven years old), she and everyone in the class knew that this meant she was *"not good at Maths,"* and that other children in the class would make mean comments about her ability because of it. This shows that even young children can cast judgement on others based on perceived mathematical ability.

Whether ability grouping is discussed in the class or not, students are often aware of being labelled as the "low group" in Maths lessons, and this can be detrimental to their relationship with Maths, and lead to feelings of anxiety and self-doubt. This becomes important in relation to whether teachers have autonomy when it comes to deciding on which students work together, or even who they sit next to in Maths lessons.

Grouping of students can be a challenging discussion and it is easy to conceptualise how it could be successful or worsen the situation when it comes to Maths Anxiety. It becomes particularly pertinent in relation to mixed ability groupings and the potential ways in which this might affect such things as peer coaching or cooperative learning. We will touch on some empirical research related to these ideas in Chapter 6.

Perhaps the most commonly reported way in which teachers can negatively impact a student's relationship with Maths, and influence the development of Maths Anxiety, is making students feel embarrassed. This may occur by asking a student, who didn't put their hand up, to answer a question in front of their classmates and they subsequently get it incorrect. This can also happen if a student is asked to come up to the front of the class and demonstrate how to solve a mathematical problem. A student may not feel confident that they can do this, and feel embarrassed that they are not able to demonstrate the required skill in front of others. Even if they know the answer, this can be anxiety-evoking for a student due to the perceived judgement from their peers and fear of making mistakes in front of others.

These feelings of embarrassment can become embedded and associated with students' perceptions of Maths, which can be detrimental and is commonly cited as a highly influential reason behind why an individual believes that their Maths Anxiety developed. It is in fact quite common for people who identify as being highly maths anxious to be able to name the teacher they believe caused their anxiety, along with the specifics of the exact situation they were in even many years later.

> Jackson and Leffingwell (1999) asked trainee teachers who had experienced Maths Anxiety or negative emotions towards Maths to describe the most negative situation they experienced when learning Maths, ranging from any time in between kindergarten and college. Just 11 participants out of 157 had only positive experiences. Out of the remaining 146: 16% of participants recalled their Maths Anxiety developing between the ages of 8-10 years, due to either difficult mathematical material, perceived hostility from their teacher, gender bias from their teacher, or feeling that their teacher was uncaring.
>
> 26% of participants recalled a negative experience between the ages of 15-17 years, due to their teacher portraying angry behaviour, unrealistic expectations, gender bias, being insensitive, or being embarrassed by their teacher in front of their peers.
> 27% of participants recalled their most negative experience during college, again in relation to their teacher's lack of sensitivity, perceived uncaring nature, poor quality of teaching, and gender bias.

Before our discussion about a teacher's influence comes to a close, it is worth looking more closely at the role of rewards and punishment in the Maths classroom, especially in relation to speed. Rewards can come in all sorts of shapes and sizes. Some obvious rewards include things such as house points, ticks, and stickers. Reward can also come in the form of praise, which can boost a student's self-esteem. It may also lead to increased feelings of panic and anxiety in maths anxious individuals though, as they could perceive any focus upon them within the Maths lesson as anxiety-evoking.

> Petronzi et al. (2017) showed how young Maths students, aged from four years old, who were mathematically confident, emphasised how they liked it when teachers acknowledged their Maths work. This could simply be when they know the teacher is peering over their shoulder as they are completing the work. On the other hand, maths anxious students described fear at teachers acknowledging incorrect Maths answers. They would talk about having to "see the teacher," which was interpreted as being a form of punishment.

There is an important distinction to make here between a teacher's intentions of a comment, and what the student perceives. A comment from the teacher, whether verbal or written, that they would like to speak to the student about their Maths work, might be intended as being supportive; that they would like to offer help and guidance. A young Maths student, or a maths anxious student, however, might see it as a stern instruction and a negative consequence of failure. Based on this, teachers should be mindful of the incongruence that can exist in such interactions. Sometimes, an explicit indication that they are offering help, rather than anything punitive, can be helpful and reassuring.

In the case of "quick Maths," such as fast completion of a Maths worksheet, this might represent a student who has grasped a mathematical concept and feels confident and perhaps even motivated to proceed through a set of Maths problems at pace. On the other hand, slower completion, as we have discussed, is often a side effect of feeling maths anxious. A maths anxious student is often aware of how they compare to their peers in terms of their Maths ability and how long they take with Maths, which, for younger students, might be perceived as pretty much the same thing. Through vicarious learning, a maths anxious student might come to associate quick Maths with reward. Conceptually, this might result in two pathways: the student fears being perceived as slow, thus feeding their anxiety and, ultimately, slower maths performance, or, they give up attempting the task so as not to run the risk of being "slow."

As such, while a Maths teacher might not overtly comment on a student's speed, they are indirectly providing feedback to them when they reward others for being "quick." That is not to say that teachers should stop rewarding Maths work that has been completed at a quick pace, but rather that there is value in being mindful of the consequences when they reward less maths anxious students in the presence of those who are high in Maths Anxiety. For instance, when discussing this with teachers, they will sometimes realise that they almost automatically praise quick completion of Maths tasks, sometimes before the content has even been checked.

> Petronzi et al. (2017) found that young students quickly come to associate failure to maintain a certain pace in Maths with adverse consequences. Typically, this involves missing out on activities that the student's peers are allowed to engage in because of having completed their work. Such activities can include going out to play at break time, and also being given the freedom to select an activity in-class, such as choosing a book to read.

This creates a battle between, on the one hand, vicariously learning that quick completion of Maths tasks leads to reward and, on the other hand, the personal experience of missing out, or even feeling isolated, due to being "slow." It is a challenging aspect of the Maths learning environment, with no clear answers.

Through their language, behaviour, and classroom culture, it is clear that teachers can have a positive influence on their students' perception of Maths, but also a negative one, leading to the potential development, or exacerbation, of Maths Anxiety. No teacher aims to negatively influence their students, and they definitely don't want their students to be anxious when faced with learning or using Maths, but it does happen, especially if they have their own negative perception of Maths (which will be looked at more in Chapter 7). If a teacher's awareness of their role in the development of Maths Anxiety is heightened, this may allow greater reflection on their classroom culture and actions, and work towards developing positive perceptions of Maths, minimising Maths Anxiety in the classroom.

Timed Situations

As previously explored, there are overlaps between the construct of Maths Anxiety and Test Anxiety, and a key feature of being tested is time pressure. Many people assume that pressure from timed situations results in poorer Maths performance among students who are highly maths anxious. This is fairly intuitive and makes sense, given that so many maths anxious students openly describe how they hate doing Maths under time pressure. However, it is worth unpacking this a little. It should be acknowledged that situations involving mathematical problems within which there is a time limit are rarely devoid of other perceived pressures. These pressures could include real or perceived consequences of poor performance, or that class tests and exams often include being observed. Therefore, the impact of simply being timed is often not tested directly.

> Szczygiel and Pieronkiewicz (2021) asked children at the beginning of elementary school why they felt anxious about Maths. One of the main reasons they reported was related to time pressure, as well as risk of failure and the fear of receiving a bad grade.

There is some empirical evidence that suggests time pressure increases the negative effect of Maths Anxiety on Maths performance, specifically when the task involves problems that tax working memory. Having said that, out of the few empirical investigations that have been carried out to test the effects of time pressure and Maths Anxiety on Maths performance on general Maths tests, most have shown little in the way of interaction between these variables. That is, time pressure usually has at least some negative impact when considered across a whole group of people, but it doesn't seem to have too much of an effect on Maths performance among those who are highly maths anxious any more than those low in Maths Anxiety.

One school of thought is that Maths Anxiety already consumes students' working memory, so additional factors that might impede cognitive processes can only have so much of an effect. This is a complex area though. Considerations need to include factors such as the nature of Maths problems and the extent to which they demand working memory resources. It might also be the case that time pressure affects learning differently to performance; this is rarely considered in the context of Maths Anxiety.

Claims that timed tests are the direct cause of the early onset of Maths Anxiety are often unsubstantiated by empirical evidence, often relying on cross-sectional studies assessing the general effect of time pressure upon Maths performance.

It is already known that there are many individual differences at play when it comes to learning Maths and its related performance, so it is likely that timed situations can impact students with different levels of anxiety in varying ways. A small amount of stress arousal, tied with the right degree of motivation, can actually result in improved Maths performance.

Therefore, for some students, time pressure might support their engagement, thinking, and ultimately performance on Maths tests, provided they are motivated to perform well.

When reflecting on the influence of time pressure, it is also important to consider what is meant by this. Is time pressure simply a time limit? Is it being told to work as quickly as you can? Or is it the case that you are encouraged to outperform your peers by being quicker than them? These are all regular features of either Maths learning or assessment. Interestingly, some research has shown that people think about time pressure during Maths tests even when time is not mentioned during the instructions, or the test itself. This tells us something about the association people make between mathematical problem-solving and time.

> **Chloe's Experience**
>
> When talking about particular situations that evoked her anxiety the most, Chloe shared that she hated when her teacher counted down the time during lessons. She said that when her teacher says *"10 minutes left,"* or *"5 minutes to go,"* Chloe can feel her heart rate increase and her chest tighten. Chloe added that this was because when she was focused on completing the task, she then felt distracted and pressured by the teacher, as well as feeling worried that she hadn't done enough to meet their expectations.

From a young age, Maths and time go hand in hand within formal schooling, to the point where many students associate being quick with being "good at Maths." It seems that by the time we are adults, although probably much sooner, there is almost an expectation that our mathematical fact retrieval and problem-solving *should* be quick. What is interesting here is what is considered "quick"; as with many behaviours, our notion of completing Maths tasks quickly is based on a comparison with how long it has taken others to perform the same task. Thus, our self-belief regarding our time-based performance might continually change based on the performance of those around us.

> Beilock and Willingham (2014) suggested that there are several reasons why alleviating time pressure might make Maths Anxiety less of a problem. They suggest it could reduce students' concerns about not finishing in time, while also giving them time and space to work through their answers.

Furthermore, expectations from teachers, and the pressure students place on themselves, can arise from performance at the group level. As with overall attainment, students and teachers can find themselves in a battle between goals that are based on individual change and those that are based on group change. If a whole Maths class works at pace, there might be a tendency to maintain, or even increase, that pace. Within that group, however, there might be individual students who are being swept along by the wider group, seemingly working at pace, but working at a rather superficial level. We should also not lose sight of the fact that time pressure might negatively impact highly maths anxious individuals from a stress and well-being perspective, more so than their low maths anxious counterparts. This is something that can often become overlooked when the focus is on mathematical understanding, the ability to retain information, and the ability of a student to perform to a particular level.

> Chinn (2009) conducted a study of several 11-13 years old in England and found that the item "having to work out answers to Maths questions quickly" was ranked as fifth out of 20 anxiety-provoking classroom experiences.

Many students find quick Maths stressful, which begs the question of why they feel this way, and whether the current approach to formal Maths learning and teaching is placing unnecessary stress on students. We should also say here that we are not downplaying the importance of speeded responses. The quick retrieval of information, such as multiplication of facts, can be highly beneficial – not just from a time perspective, but also from the point of view of reducing working memory resources required when solving a Maths problem.

Also, at times, the real world does require mathematical problem-solving to be at a particular pace, and we need to prepare students for this. However, there are many instances in which Maths does not have to be quick in the real world and, indeed, taking a breath and slowing down could actually be more productive. It should also be noted that the fear associated with quick Maths could trigger a downward trajectory in a student's general anxiety, attitudes, and motivation associated with learning and using Maths. As such, it is always worth considering whether it is counterproductive to encourage Maths learning at speed before a student feels sufficiently confident.

> Chinn (2012) highlighted that the culture within schools to promote the need for speed and accuracy is in direct conflict with the need to develop students' confidence and self-belief.

It is now commonplace to use online digital platforms as a method for mathematical learning and assessment, and it is interesting to hear about the different ways that schools use them. They nearly always involve the monitoring of a student's speed, often presenting back to them how long they have taken. In many cases, students are informed of where they appear in the hierarchy of response times; this could be relative to their classmates or even among all users of the platform. Many students enjoy this competitive aspect of learning, especially if it gives them an opportunity to demonstrate to themselves, and others, the comparatively quick Maths they can do. When it comes to students with a particularly high level of Maths Anxiety, however, such platforms can be a threat to their mathematical self-belief, providing frequent reinforcement of a negative self-belief concerning a student's "slowness." In some cases, a hierarchy of response times is made publicly available to groups of students. Presumably, the intention is to motivate those at the bottom of the hierarchy to increase their efforts. Of course, this is based on the assumption that an increase in effort is needed.

While it can be said that timed situations influence the experience of Maths Anxiety, it is clear that there are many other factors to consider. The need for rote memorisation and speed of recall, the comparison to others, and the fear of failure and negative consequences can all influence the anxiety experienced within timed situations involving the use of Maths.

Specific Mathematical Concepts

It is unlikely that an individual's feelings of anxiety are running at 100% all the time. As we have explored, there may be times when a certain mathematical situation or environment intensifies a maths anxious individual's feelings. Similarly, there may be mathematical situations that don't evoke the same amount of anxiety as others, leading to a feeling of comparative comfort. When using or learning Maths, individuals need to engage with a range of mathematical concepts, which vary in their requirements. For example, one particular concept might require extensive prior knowledge, or place a lot of pressure on working memory, or it might be based on a completely new concept without relatable prior learning to help make connections. Alternatively, a concept may be quite familiar to an individual, place little pressure on their working memory, and allow them to draw on their previous success and understanding.

A maths anxious person may also recall previous negative experiences when learning a specific concept, such as finding it difficult to successfully use short division for example or feeling embarrassed when answering a question about ratio incorrectly in front of the class. This may influence their experience of Maths Anxiety and negatively impact their self-belief and attitude towards Maths. Or, an individual may recall experiencing success when previously learning and applying a mathematical concept, for example, being praised by the teacher for their understanding of the place value of decimal numbers, or finding it relatively straightforward to use formal methods for multiplication accurately. This may lead to increased feelings of comfort when working with these specific concepts.

At present, there isn't a wealth of research into how different mathematical concepts vary in their influence upon Maths Anxiety, but it is an influence that we see as important to discuss, particularly for teachers. By looking at different mathematical concepts in detail, we can better understand if there are any particular areas that are more likely to influence the experience of Maths Anxiety, whether through exacerbating negative responses or creating feelings of comfort.

The Number System

Understanding how to read and write numbers, and know what the written form is representing, is key to successfully using Maths. Termed "place value," students develop their understanding of the number system through initially learning what the value of each digit within these numbers represents, and across various representations. As a concept, this does not usually elicit strong feelings of anxiety, as it does not require manipulation of numbers and is based on clear parameters that remain constant, as there are always only 10 digits to represent numbers available, for example. Often, teachers discuss it as something that doesn't necessarily become harder if bigger numbers are used. For example, if students are identifying the place value of a 4-digit number, it's not largely different to identifying the place value of a 5-digit number, as they are both integers and follow the same rules of the number system. As such, learning about the place value of numbers may be considered as a comfort zone for some, especially as this is something that's regularly revisited throughout a student's education and is therefore based on developing prior understanding.

But, what happens when numbers start to deviate from the rules associated with integers that we are taught when we are younger, and are no longer whole?

Decimal Numbers

Exploring the place value of decimal numbers can be emotionally unsettling if a student is uncertain about the concept of numbers which are not whole. In the first instance, students are taught that, beginning from 0, there are ones, tens, hundreds, and so on, creating a stable depiction of what a number is in their mind. Years later, however, they are taught that numbers can be smaller than 1, and they appear after a decimal point to the right of the ones column – the column which they have thought to be the lowest value. This can be quite challenging for students, particularly if they viewed place value as a mathematical area where they felt comfort. Even the appearance of a decimal point can cause a negative response, particularly if a maths anxious student feels uncomfortable working with unfamiliar numbers and signs.

Fractions

Fractions are taught much earlier on than decimals in the Maths curriculum, but often in relation to shapes or objects. Even when they are linked to numbers, fractions are often illustrated as part of a larger number, so the fraction itself will be an integer as well, such as $1/3$ of 21 being 7. This may make it difficult for students to associate fractions as being less than a whole number, and conceptually relate them to the number system. This may link to how, as a concept, fractions are often discussed negatively, and can be highly anxiety-evoking, rather than an area of comfort.

> **Ollie's Experience**
>
> When asked whether there were any mathematical concepts that made him feel the most anxious, Ollie quickly said *"fractions, because they are ridiculously hard."* Exploring this further, Ollie shared that he never understood them, unless they were visual fractions, and even then, he had to try really hard to stay focused and not let his anxiety take over. One thing in particular he disliked was that they could be the same as other fractions (referring to equivalent fractions) and that this process was very time-consuming and challenging.

Students often share their dislike of fractions, without being able to give a reason behind this. Fractions are seemingly socially accepted as being difficult, which may impact an individual's perception of the concept, before they have truly experienced it themselves. Understanding fractions as a concept can be challenging, as on top of being represented as shapes, objects, numbers, decimals, and percentages, they can be presented as mixed numbers or improper fractions, and they can be equivalent to other fractions, and also simplified. The language associated with fractions can also be daunting, with terms such as numerator, denominator, and equivalent, to name a few. This can create a barrier to a person understanding what someone is referring to, and making learning and using fractions more challenging, increasing the potential for anxiety surrounding it.

All of this occurs before fractions are even manipulated. There are many rules and procedures that can be followed in relation to this. For example, a student may be taught that to divide fractions, all they need to do is swap the numerator and denominator round in the second fraction, and then multiply both fractions' numerators together, then both fractions' denominators together. This has multiple steps to follow but does require only procedural knowledge, rather than conceptual, which may alleviate some pressure. However, teamed with various other ways in which to manipulate fractions and solve various calculations, these can be quite daunting to recall and utilise in order to successfully use fractions across multiple situations.

> Hunt et al. (2019) suggested that Maths Anxiety can be specific to more abstract forms of Maths.

Percentages

Alongside fractions and decimals, percentages are taught as an equivalent to these and, conceptually, as part of an amount or number. Being associated with fractions may perhaps increase the anxiety surrounding the concept of percentages, but as percentages are used most in everyday life, this can provide an appreciation of their utility and perhaps a level of familiarity or comfort to some students. The difficulty associated with percentages can differ widely as well, as students may find it straightforward to find 50% or 10% of an amount due to knowing their 10 times tables or being able to halve a number. However, finding 27% of a number may prove challenging, with more steps needed, or even finding 50% of an odd number or of a decimal. Again, how an individual reacts to using and learning decimals may relate to their understanding of number and their relationship with similar concepts.

Ratio

When introduced as a new topic, ratio can be anxiety-inducing as it is unfamiliar. Ratio is a similar yet contrasting concept to a student's previous learning of fractions, as fractions are parts of a whole and can be identified by finding equal parts, whereas ratio is instead a comparison of parts. For example, a ratio of 1:2 means there are 3 whole parts within the whole, whereas a fraction of ½ means that 1 whole is split into 2 equal parts. There may be some conceptual confusion with this, due to the similarity with fractions but distinct differences, leading to anxiety when dealing with this concept and successfully using ratio. When ratio is initially introduced into the curriculum though, the numbers used are quite small and accessible, usually in the single-digit range, which may seem quite achievable and comfortable for some individuals.

Algebra

> **Ryan's Experience**
>
> Ryan hadn't even learnt algebra, as he was only in Year 4 at the time, but he knew of it. He spoke about how it's the hardest thing you can do in Maths and *"even professors like Einstein find it hard."* This highlights the role that social perceptions can have upon particular mathematical concepts.

As the number system is mostly represented numerically in a student's early Maths education, the concept of algebra can be quite overwhelming when introduced. Algebra is often spoken about as a complex concept that is difficult to understand, and one that causes intense negative emotions. Whether this is due to the substitution of letters into Maths, which have previously been isolated to other subjects, or looking at the concept of numbers in a different light, many people have recollections of negative experiences of learning and using Algebra. This is often seen as anxiety-evoking, and a novel concept when introduced, even though there is prior learning that can be drawn upon.

> Ferguson (1986) explored the notion of "Abstraction Anxiety," highlighting that anxiety was more likely to occur towards abstract mathematical concepts, such as algebra.

The Four Operations

Alongside their understanding of number, students learn and use four key mathematical concepts in order to manipulate numbers: addition, subtraction, multiplication, and division.

Addition

This is quite a varied mathematical concept, which can elicit anxiety but also provide comfort to maths anxious individuals. Students may be taught various ways in which to add numbers together, such as partitioning, compensating, using number lines, base 10, or 100 squares, as a few examples. Providing children with various ways in which to learn the skill of addition is useful, as there is more chance to have a positive impact on a student's understanding by providing different ways to succeed, in the hope that they find one they like. However, this can be quite overwhelming and may influence the experience of Maths Anxiety. Being faced with multiple options and choices to make can be difficult, leading to a negative perception of the task and addition more generally.

This is heightened if the addition of numbers involves crossing boundaries, where some learnt rules don't always work. For example, if a child is taught to partition numbers to add, they may partition 82 + 19 into 80 + 10, 90, and then 2 + 9, 11. Now the student is faced with a dilemma or confusion, as they can't merge this back together. If they do, it could lead to an answer of 9011, solidifying misconceptions or making mistakes. If a student cannot rely on an addition rule always working in a consistent manner, this can cause anxiety.

Alternatively, a way in which addition can often be seen as a comfort zone by a maths anxious individual is when using the formal written method to add numbers together. By aligning numbers and working through each column in turn, this turns a concept into a procedure, potentially removing any anxiety felt, even when carrying over into the next column. This requires minimal use of an individual's working memory and doesn't rely on a solid mathematical understanding of the concept. Interestingly, multi-digit addition problems are commonplace in Maths Anxiety research that adopts an experimental approach. They provide a way for researchers to control for all sorts of features of Maths problems that could impact performance, including problem size. They also provide a convenient way of presenting problems based on the existence (or not) of a carry operation, to manipulate the extent to which working memory resources are required.

Amber's Experience

Amber identified that using the formal written method for addition was quite easy, and she didn't feel anxious doing this at all. When asked why she thought this was the case, she said that all you need to know is adding two single digit numbers together, which *"you get taught when you're really young."* This highlights how procedural methods might cause some relief in maths anxious students but doesn't necessarily support progressing their mathematical understanding.

Subtraction

This often evokes more anxiety than addition. Subtraction requires a solid understanding of numbers and how to manipulate them, as well as a good working memory to retain information and hold numbers while taking away. It can be particularly anxiety-evoking when borrowing from other place value columns, as individuals again need a solid understanding of number. This can be heightened when mental subtraction is being used, which places pressure on a person's working memory, and also relies on their ability to manipulate a range of numbers without support.

Like addition, students may be taught multiple ways in which to solve subtraction calculations, which can result in confusion and feelings of panic, particularly if these methods cannot be transferred across multiple situations. Being taught how to subtract using number lines, Base 10, 100 squares, partitioning, counting on or counting back, or compensating, can again provide lots of support to a variety of students but can also be overwhelming and disconcerting.

Comfort can be provided through the procedure of formal written subtraction, similar to addition, as an individual does not need to have an in-depth grasp of the concept. Rather, they can learn the steps of the procedure and work to solve calculations using this technique, without exerting pressure on their working memory or working with large numbers all at once. It is useful to remember that correctly recalling and implementing a set of procedures can boost a student's confidence in the moment, which might facilitate greater engagement later on. As discussed though, an over-reliance on procedural knowledge and practice is not effective in the long-run.

> Kirkland (2020) found that the use of written methods for addition and subtraction was viewed as areas of comfort for all three students participating in the case study research.

Multiplication

Recalling multiplication facts is often one of the pressures felt early on in an individual's mathematical journey. In England, for example, it is an expectation that a student can fluently recall the multiplication facts up to 12 x 12 by the time they are nine years old, which is measured nationally by the Multiplication Tables Check (MTC). These key facts are often learnt through rote memorisation and continual practice, which can be anxiety-evoking and overwhelming. This is heightened even more when a student needs to apply this and manipulate these known multiplication facts to problem solve, such as "How many eggs are in 32 full boxes, if each box can hold 8 eggs?"

> **Lara's Experience**
>
> Lara felt that she was ok using written multiplication, such as short multiplication and long multiplication, and that they didn't cause her much anxiety. But when she was asked to recall multiplication facts quickly, she felt that her mind went blank and she couldn't answer any questions at all.

With various stages of multiplication using integers, such as multiplying a 1-digit number by another 1-digit number, then by a 2-digit number, and so on, there are many strategies that children are taught. Ranging from partitioning, the grid method, compensating, rounding, and arrays, this again can feel overpowering, especially if all of these strategies do not feel understandable to a student.

The written formal method for multiplication, also referred to as "short multiplication," can again provide a way in which an individual may reduce their anxiety and feel more comfortable using the multiplication skills required, particularly if this is multiplying by a 1-digit number. For example, multiplying 364 by 6 requires a student to multiply the 4 by the 6, then the 6, then the 3, with little reflection on the conceptual meaning. This does become harder when it's multiplying by a 2-digit number, or more (using the process called "long multiplication"), due to more processes being required, which may lead to anxiety. Regardless, using any written method for multiplication does require the individual to have good recall of multiplication facts.

> Helmane (2016) researched 124 students' positive and negative emotions towards specific mathematical concepts in Grade 3 across 5 schools in Latvia. Using questionnaires, they found that the emotions felt towards various mathematical concepts differed:
>
> **Addition and Subtraction:** 71% felt positive emotions; 13% felt negative emotions.
> **Multiplication:** 22% felt positive emotions; 74% felt negative emotions.
> **Fractions:** 34% felt positive emotions; 61% felt negative emotions.

Division

Alongside multiplication facts, students in England are expected to know their division facts associated with the 12 times tables and below by the time they are 9–10 years old. While not tested in the MTC like multiplication facts are, it is an expectation set out within England's National Curriculum. Looking at the inverse of multiplication facts is a way in which this learnt fluency is encouraged, but it again results from rote memorisation and successful recall of multiplication facts. If an individual finds it difficult to identify the associated multiplication facts, being asked to fluently recall division facts can be anxiety-evoking, and being asked to recall these at speed can add further pressure and feelings of panic. Anxiety can also occur when being asked to apply these facts and solve problems, such as "A lunch box can hold 3 sandwiches. How many lunch boxes are needed for 24 sandwiches?". If a student does not feel comfortable knowing these facts fluently, and if they do not have any visual aids such as a multiplication chart to help them, this can influence feelings of anxiety. Moreover, word problems such as this involve a further level of linguistic complexity, which has only recently started to gain attention in the context of Maths Anxiety.

Again, being taught numerous methods such as using arrays to share or using repeated subtraction known as chunking, might feed into anxiety because of feeling overwhelmed. It is often beneficial to use the method of short division, also known as the "bus stop method." This formal method can create comfort, as again, it's procedural in nature. By knowing how to divide each digit in turn, and carrying over to the next digit in the number, individuals can follow this procedure without taxing working memory and needing to securely know a wide range of division facts.

The application of short division only applies to dividing by single digits however, as when dividing by a 2-digit number, or more, individuals are taught the method of long division, which has long been heralded as an extremely tricky concept to master. While it is a procedure, it's a procedure involving multiple steps for each digit (divide, multiply, subtract, bring down). This can lead to feelings of anxiety as it is a complex process and requires the use of working memory in order for this to become a successful way in which to divide. Perhaps as it is socially identified as a tricky concept to master, individuals may also have a preconception that it is challenging, again contributing to the experience of Maths Anxiety.

Liam's Experience

Liam really disliked division, particularly long division. He shared that he learnt long division a lot in primary school, and that they always went over and over it, but he never understood it. Liam said that they don't use it now, because they just use calculators, but sometimes they need to know division facts which does cause him some anxiety. When asked to explain this further, he said it makes him feel anxious because he feels that he needs to understand his multiplication facts first, to then know the opposite in division.

This can be heightened further when remainders are involved. For whichever method is used, remainders add another layer of complexity to division, as students are initially taught at a young age that when dividing, things can be shared nicely. This then leads to the potential of having something left over, such as a sweet or a pencil, but when this is put in abstract terms with a number, and the end of the dividend has been reached, students can become overwhelmed and feel let down by the procedure used. As students become older, the options of having remainders as a fraction, as a decimal, or as a standalone remainder again add further complexity, which can increase anxiety due to the uncertainty and lack of clarity.

Statistics

This concept is often seen as a comfort zone for younger students, who are introduced to data and statistics primarily in the form of tally charts, pictograms, and bar charts. Being able to collect their own data can help form a relationship with, and understanding of, the data and how it is represented. Visual representations of this can be comforting to some students, yet others who are confident with numbers may find it challenging to convert from numbers to pictorial representations. Most students enjoy this though and can work with numbers represented in different forms without evoking feelings of anxiety. In extreme cases of Maths Anxiety, however, even data in non-numerical representations can be intimidating, perhaps because of the underlying numerical aspects.

As visual representation of data becomes more complex and looks at changes in numbers and progression instead of constant numbers, such as line graphs to measure changes in speed for example, this can strain a student's individual conceptual understanding of data and how it is represented. Rather than being stable representations, the data is now representative of change: a more abstract concept. This change in conceptual understanding can lead to uncertainty in students, and anxiety.

Describing data and making inferences within statistics can also be anxiety-evoking. For example, having to recall how to calculate the central tendency (the mean, the median, and the mode) can be demanding, particularly if the numbers involved are complex or bountiful. The fact that these three measures all begin with the same phonetic sound as well often plays into the confusion of which word refers to which statistical measure. This demand, and associated anxiety, can increase as different statistical measurements, such as measures of dispersion and variation (standard deviation, for example), are required to identify the meaning behind what the numbers are representing.

There is an entire body of research that focuses on Statistics Anxiety, including specific self-report scales to measure it. There has also been a debate in the academic literature concerning the conceptual overlap between Statistics Anxiety and Maths Anxiety, which is beyond the scope of this book. Suffice to say, Statistics Anxiety has mostly been studied in university students, in which statistics learning is included as part of their degree programme.

Geometry

> **Jack's Experience**
>
> Jack weighed up how he felt about specific mathematical content, and, while manipulating numbers made him feel anxious, he shared that he *"didn't mind Maths when it's about shapes."* He found it difficult to expand on this more, but he did explain that he liked being able *"to see the Maths in front of him,"* and that he felt that no one was significantly better than anyone else when learning about shapes, compared to the differences he spotted between people doing calculations or using fractions, for example.

A concept that is more visual than number, exploring shapes and their geometrical properties can often be an area of Maths that provides comfort to students, rather than prompting feelings of anxiety. Whether this is because geometry is primarily quite visual in nature, or seen in everyday life on a daily basis, individuals don't often feel overwhelmingly anxious when dealing with shapes and their properties. Individuals often identify learning about geometry as a separate entity in Maths, as the concept is not based solely upon understanding and manipulating numbers. Interestingly, from Heidi's teaching experience, she has identified that those students who are often extremely confident with numbers are less certain about geometry, and vice versa for those who are less confident with number. Although this may not be true for everyone, this does highlight how Maths as a whole may not have the same impact on Maths Anxiety as a constant.

> Hunt et al.'s (2019) Mathematics Calculation Anxiety Scale incorporates geometry questions, which appear to combine with algebra questions as part of the measurement of Abstract Maths Anxiety.

As students' progress through their schooling, using and learning geometry begins to include more complex concepts that are not always relatable, involving the study of angles, distance, proportion, dimension and position, and theories such as the Pythagorean Theorem. This requires students to be comfortable with number, algebra, and formulae, which can contribute to the experience of Maths Anxiety if an individual is not.

> Vukovic et al. (2013) conducted a longitudinal study of 113 students moving from the second to the third grade in the US. They found that individual differences in students' calculation skills and mathematical applications could be explained by Maths Anxiety, but Maths Anxiety did not explain individual differences in children's geometric reasoning.

Measurement

Measurement often encompasses a range of concepts within it, such as mass, capacity, and length as a few examples, so there are a few ways in which learning and using measurement may evoke anxiety and influence the experience of Maths Anxiety. One aspect of measurement that may be overwhelming for individuals is how one concept can be represented in various ways, such as on a vertical or horizontal measurement line, or digital or analogue scales, for example. In addition to this, the scales of measurement used can vary in how much each interval is worth. For example, one weighing scale may have intervals of 500 grams represented by each line, whereas another may have intervals of 100 grams. Having this lack of consistency can be overwhelming and evoke feelings of panic and anxiety. Similarly, there are various, and differing, formulas for converting between different measurements, such as 1000 metres being in 1 kilometre, but 100 centimetres being in 1 metre, and 10 millimetres in 1 cm. On top of this, an individual may have to convert between imperial and metric measurements, which is even more complex and can be worrisome.

Money

Measurement is inclusive of the concept of money as well, which has high real-world applicability. Money is a concept that is engaged with from an early age, but with the necessity of needing to understand numbers, and how to manipulate them, it can result in the experience of Maths Anxiety both in and out of the classroom. Managing money, exchanging money, converting to different currencies, and understanding the concept of money as a whole can create feelings of anxiety in individuals, with perceived pressure in ensuring money management is successful due to the negative consequences associated with it if it isn't.

The complexity of having a range of coins, paper notes, and credit and debit cards can also create anxiety in understanding money as a concept, surrounded by feelings of uncertainty. In a world that is becoming increasingly digitised, physical manipulation of money is experienced less often. This poses some questions around the usefulness of actual coins and paper money as learning tools. Does the physical manipulation of cash support confidence with numbers, or does it potentially introduce unfamiliarity now that many real-world payments are done digitally? Such questions will be at the forefront of many Maths teachers' minds, but they become especially relevant in the case of students with Maths Anxiety.

Time

Another form of measurement that can lead to the development, or exacerbation of, Maths Anxiety, is the concept of time. Teachers often explain to students that it's a concept that takes time to understand and master. But in the meantime, this can cause feelings of uneasiness and anxiety. As the concept of time is a continuous loop, this can evoke anxiety as it differs from other linear forms of measurement. Time presented in digital format is often less anxiety-evoking for students, as this does not require a deep conceptual understanding in the way that analogue time does. However, learning and understanding the concept of time involves the use of both digital and analogue time, which can make it difficult for an individual if they rely on the digital format only.

There are also different units of time as well, which can be quite hard to liken to each other. For example, there are 60 seconds in 1 minute, and 60 minutes in 1 hour, but then there are 24 hours in a day, and 7 days in a week, with 4 weeks in a month, but 12 months in a year. This again creates pressure on students to recall these facts to successfully understand the concept of time, which is a part of everyday life as well. This can create feelings of anxiety, as again, the consequences of not being able to do this can negatively impact everyday life.

Time in particular is something that many people with Dyscalculia struggle with, possibly due to difficulties with number sense. Given the constant relevance of time in everyday life, it is no wonder that anxiety can be high as soon as time is mentioned, but especially high if someone has Dyscalculia. The notion of a combined effect of Maths Anxiety and Dyscalculia is something that we have not explored here, but it is worth highlighting as an area that requires attention, both in terms of research and practice.

> ### Carl's Experience
>
> One thing that Carl found particularly anxiety-evoking was telling the time. He recalled that his teacher always used to tell the class that *"once you've got it, you've got it,"* but he always thought *"what if I never get it?"* He still finds telling the time to be overwhelming if he is in a rush, or asked on the spot by someone what the time is and he doesn't have a digital device to hand.

Looking at the specific mathematical concepts that are learnt and used, while not exhaustive, does show how some concepts can influence the experience of Maths Anxiety, while some areas may not elicit anxiety and instead might create feelings of calm and comfort. These areas of comfort are specific to each individual, perhaps linked to their experiences and understanding of each mathematical concept. Often, however, concepts and methods that are procedural in nature elicit less anxiety than those that require a deeper understanding. These are likely to elicit less anxiety because they may involve easy to follow methods, without placing too much pressure on an individual's working memory, or the need to understand the mathematical concept behind what is happening.

This may indicate that Maths Anxiety is not a stable phenomenon and may be influenced by what specific mathematical concept is being used, such as simple geometry or place value, compared to those which are more likely to elicit a higher anxiety response, such as fractions or algebra. This highlights how important it is to reflect on which situations evoke anxiety for an individual, both in and outside of the classroom, to better understand their experience of Maths Anxiety.

So... Who and What Can Influence It?

- Many factors can influence the experience of Maths Anxiety, whether positively or negatively.
- Other people can influence Maths Anxiety, such as parents, siblings, peers, and teachers.
- Timed situations can increase an individual's feelings of panic and anxiety.
- Some mathematical concepts can negatively influence Maths Anxiety, primarily those which are more abstract and require a strong conceptual understanding.
- There are mathematical concepts and calculation methods which may be less likely to evoke anxiety, which are predominantly visual or procedural in nature.
- Each individual is unique, and what influences Maths Anxiety can differ greatly between people.

Influences upon Maths Anxiety vary for each individual, but by reflecting on those that are well-researched, or those that are regularly reported, a few important trends may emerge. The role of other people is quite significant, as how an individual views Maths, interacts with it, and views their relationship with it, can be moulded by those around them. Teachers and peers can negatively, and positively, influence Maths Anxiety primarily in the classroom, while siblings and parents may be influential primarily in the home. Comparison to others is key here, as how a person views how others engage with Maths, and perceives their success and enjoyment in relation to their own, can have a highly negative impact.

Mathematical situations, including evaluative contexts such as those that are timed, or situations that are related to certain mathematical concepts, can again influence a person's experience of Maths Anxiety. Being timed can intensify an individual's anxiety and pressure felt to succeed, while learning and using mathematical concepts that require a deep understanding in order to manipulate numbers or utilise abstract theory, such as algebra or fractions, can contribute to an individual's experience of Maths Anxiety as well.

What is important to reflect upon here though is that these influences are also interlinked with various other factors involved in the experience of Maths Anxiety. For example, feeling embarrassed in front of others and comparing mathematical ability is related to an individual's attainment and self-belief, while learning specific mathematical concepts is related to these factors too, as well as attitudes towards Maths, and working memory. This further shows how Maths Anxiety is a complex phenomenon with many interlinked factors that can be influential when considering the development, maintenance, and exacerbation of Maths Anxiety.

References

Ashcraft, M. H. and Krause, J. A. (2007) 'Working Memory, Math Performance, and Math Anxiety', *Psychonomic Bulletin & Review*, 14, pp. 243–248. Available at: https://doi.org/10.3758/BF03194059.

Bartley, S. R. and Ingram, N. (2017) 'Parental Modelling of Mathematical Affect: Self-Efficacy and Emotional Arousal', *Mathematics Education Research Journal*, 30, pp. 277–297. Available at: https://dx.doi.org/10.1007/s13394-017-0233-3.

Becker, M., Litkowski, E. C., Duncan, R. J., Schmitt, S. A., Elicker, J. and Purpura, D. J. (2022) 'Parents' Math Anxiety and Mathematics Performance of Pre-Kindergarten Children', *Journal of Experimental Child Psychology*, 214, p. 105302. Available at: https://doi.org/10.1016/j.jecp.2021.105302.

Beilock, S. L. and Willingham, D. T. (2014) 'Math Anxiety: Can Teachers Help Students Reduce It? Ask the Cognitive Scientist', *American Educator*, 38(2), p. 28.

Berkowitz, T., Gibson, D. J. and Levine, S. C. (2021) 'Parent Math Anxiety Predicts Early Number Talk', *Journal of Cognition and Development*, 22(4), pp. 523–536. Available at: https://doi.org/10.1080%2F15248372.2021.1926252.

Bokhorst, C. C., Sunter, S. R. and Westenberg, P. M. (2010) 'Social Support From Parents, Friends, Classmates, and Teachers in Children and Adolescents Aged 9 to 18 years: Who is Perceived as Most Supportive?', *Social Development*, 19, pp. 417–426. Available at: https://dx.doi.org/10.1111/j.1467-9507.2009.00540.x.

Chinn, S. (2009) 'Mathematics Anxiety in Secondary Students in England', *Journal of Dyslexia*, 15(1), pp. 61–68. Available at: https://doi.org/10.1002/dys.381.

Chinn, S. (2012) 'Beliefs, Anxiety, and Avoiding Failure in Mathematics', *Child Development Research*, 20(12), pp. 1–8. Available at: https://doi.org/10.1155/2012/396071.

Daucourt, M.C., Napoli, A. R., Quinn, J. M., Wood, S. G. and Hart, S. A. (2021) 'The Home Math Environment and Math Achievement: A Meta-Analysis', *Psychological Bulletin*, 147(6), p. 565.

DiStefano, M., Retanal, F., Bureau, J. F., Hunt, T. E., Lafay, A., Osana, H. P., Skwarchuk, S. L., Trepiak, P., Xu, C., LeFevre, J. A. and Maloney, E. A. (2023) 'Relations Between Math Achievement, Math Anxiety, and the Quality of Parent–Child Interactions While Solving Math Problems', *Education Sciences*, 13(3), p. 307. Available at: https://doi.org/10.3390/educsci13030307.

Ferguson, R. D. (1986) 'Abstraction Anxiety: A Factor of Mathematics Anxiety', *Journal for Research in Mathematics Education*, 17(2), pp. 145–150. Available at: https://doi.org/10.2307/749260.

Finlayson, M. (2014) 'Addressing Math Anxiety in the Classroom', *Improving Schools*, 17(1), pp. 99–115. Available at: https://doi.org/10.1177/1365480214521457.

Fiore, G. (1999) 'Math-Abused Students: Are We Prepared to Teach Them?', *The Mathematics Teacher*, 92(5), pp. 403–406.

Foley, A. E., Herts, J. B., Borgonovi, F., Guerriero, S., Levine, S. C. and Beilock, S. L. (2017) 'The Math Anxiety-Performance Link: A Global Phenomenon', *Current Directions in Psychological Science*, 26(1), pp. 52–58. Available at: https://doi.org/10.1177/0963721416672463.

Garba, A., Ismail, N., Osman, S. and Rameli, M. R. M. (2020) 'Exploring Peer Effect on Mathematics Anxiety Among Secondary School Students of Sokoto State, Nigeria Through Photovoice Approach', *Eurasia Journal of Mathematics, Science and Technology Education*, 16(2), pp. 1–12. Available at: https://doi.org/10.29333/ejmste/112622.

Gest, S. D. and Rodkin, P. C. (2011) 'Teaching Practices and Elementary Classroom Peer Ecologies', *Journal of Applied Developmental Psychology*, 32(5), pp. 288–296. Available at: https://doi.org/10.1016/j.appdev.2011.02.004.

Good, C., Rattan, A. and Dweck, C. S. (2012) 'Why Do Women Opt Out? Sense of Belonging and Women's Representation in Mathematics', *Journal of Personality and Social Psychology*, 102(4), p. 700. Available at: https://doi.org/10.1037/a0026659.

Harper, N. W. and Daane, C. J. (1998) 'Causes and Reduction of Math Anxiety in Preservice Elementary Teachers', *Action in Teacher Education*, 19(4), pp. 29-38. Available at: https://doi.org/10.1080/01626620.1998.10462889.

Hatfield, E., Cacioppo, J. T. and Rapson, R. L. (1992) 'Primitive Emotional Contagion', in M. S. Clark (ed.) *Emotion and Social Behaviour*. London: Sage Publications, pp. 151-177.

Helmane, I. (2016) 'Emotions of Primary School Pupils in Mathematics Lessons', *Signum Temporis*, 8(1), p. 22. Available at: https://dx.doi.org/10.1515/sigtem-2016-0013.

Hunt, T. E., Bagdasar, O., Sheffield, D. and Schofield, M. B. (2019) 'Assessing Domain Specificity in the Measurement of Mathematics Calculation Anxiety', *Education Research International*, 2019, pp. 1-7. Available at: https://doi.org/10.1155/2019/7412193.

Ismail, N., Garba, A., Osman, S., Ibrahim, N. H. and Bunyamin, M. A. H. (2022) 'Exploring Teacher Effects on Intensifying and Minimizing Mathematics Anxiety among Students in Sokoto State, Nigeria', *International Journal of Evaluation and Research in Education*, 11(1), pp. 161-171. Available at: https://dx.doi.org/10.11591/ijere.v11i1.22189.

Jackson, C. D. and Leffingwell, R. J. (1999) 'The Role of Instructors in Creating Math Anxiety in Students from Kindergarten Through College', *The Mathematics Teacher*, 92(7), pp. 583-586. Available at: https://doi.org/10.5951/MT.92.7.0583.

Jameson, M. M. (2014) 'Contextual Factors Related to Math Anxiety in Second-Grade Children', *The Journal of Experimental Education*, 82(4), pp. 518-536. Available at: https://dx.doi.org/10.1080/00220973.2013.813367.

Kim, J., Shin, Y. J. and Park, D. (2023) 'Peer Network in Math Anxiety: A Longitudinal Social Network Approach', *Journal of Experimental Child Psychology*, 232, p.105672. Available at: https://doi.org/10.1016/j.jecp.2023.105672.

Kirkland, H. R. (2020) *An Exploration of Maths Anxiety and Interventions in the Primary Classroom*. Ph. D. Thesis. University of Leicester.

Lane, C. (2017) 'Student's Images of Mathematics: The Role of Parent's Occupation', in Dooley, T. and Gueudet, G. (eds.) *Proceedings of the Tenth Congress of the European Society for Research in Mathematics Education (CERME10)*. Dublin: ERME, pp. 1130-1337.

Maloney, E. A., Ramirez, G., Gunderson, E. A., Levine, S. C. and Beilock, S. L. (2015) 'Intergenerational Effects of Parents' Math Anxiety on Children's Math Achievement and Anxiety', *Psychological Science*, 26, pp. 1480-1488. Available at: https://psycnet.apa.org/doi/10.1177/0956797615592630.

OECD (2015) *Does Math Make You Anxious? PISA in Focus, 48*. Paris: OECD Publishing. Available at: https://doi.org/10.1787/5js6b2579tnx-en.

O'Toole, T. and Plummer, C. (2004) 'Social Interaction: A Vehicle for Building Meaning', *Australian Primary Mathematics Classroom*, 9(4), pp. 39-42.

Petronzi, D., Staples, P., Sheffield, D., Hunt, T. and Fitton-Wilde, S. (2017) 'Numeracy Apprehension in Young Children: Insights from Children Aged 4-7 Years and Primary Care Providers', *Psychology and Education*, 54(1), pp. 1-26.

Sari, M. and Hunt, T. E. (2020) 'Parent-Child Maths Affect as Predictors of Children's Mathematics Achievement', *International Online Journal of Primary Education*, 9, pp. 85-96.

Silver, A. M., Elliott, L. and Libertus, M. E. (2021) 'When Beliefs Matter Most: Examining Children's Math Achievement in the Context of Parental Math Anxiety', *Journal of Experimental Child Psychology*, 201, p. 104992. Available at: https://doi.org/10.1016/j.jecp.2020.104992.

Soni, A. and Kumari, S. (2015) 'The Role of Parental Math Anxiety and Math Attitude in their Children's Math Achievement', *International Journal of Science and Mathematics Education*, 15, pp. 331-347. Available at: https://dx.doi.org/10.1007/s10763-015-9687-5.

Szczygieł, M. and Pieronkiewicz, B. (2021) 'Exploring the Nature of Math Anxiety in Young Children: Intensity, Prevalence, Reasons', *Mathematical Thinking and Learning*, 24(3), pp. 248-266. Available at: https://doi.org/10.1080/10986065.2021.1882363.

Szucs, D. and Toffalini, E. (2023) 'Maths Anxiety and Subjective Perception of Control, Value and Success Expectancy in Mathematics', *Royal Society Open Science*, 10(11), p. 231000. Available at: https://doi.org/10.1098/rsos.231000.

van der Vleuten, M., Weesie, J. and Maas, I. (2020) 'Sibling Influence in Field of Study Choices', *Research in Social Stratification and Mobility*, 68, p. 100525. Available at: https://doi.org/10.1016/j.rssm.2020.100525.

Vukovic, R. K., Kieffer, M. J., Bailey, S. P. and Harari, R. R. (2013) 'Mathematics Anxiety in Young Children: Concurrent and Longitudinal Associations with Mathematical Performance', *Contemporary Educational Psychology*, 38(1), pp. 1-10. Available at: https://doi.org/10.1016/j.cedpsych.2012.09.001.

Wang, Z., Hart, S. A., Kovas, Y., Lukowski, S., Soden, B., Thompson, L. A., Plomin, R., McLoughlin, G., Bartlett, C. W., Lyons, I. M. and Petrill, S. A. (2014) 'Who's Afraid of Math? Two Sources of Genetic Variance for Mathematical Anxiety', *Journal of Child Psychology and Psychiatry*, 55(9), pp. 1056-1064. Available at: https://doi.org/10.1111/jcpp.12224.

Yanuarto, W. N. (2016) 'Teachers Awareness of Students' Anxiety in Math Classroom: Teachers' Treatment VS Students' Anxiety', *Journal of Education and Learning (EduLearn)*, 10(3), pp. 235-243. Available at: https://doi.org/10.11591/edulearn.v10i3.3808.

6   What Can We Do?

People often ask us how they can help someone who is experiencing Maths Anxiety; what can they do to fix the problem? It's quite a difficult question to answer, as it depends on so many factors. As we have seen, each maths anxious individual experiences Maths Anxiety in different ways. Therefore, it stands to reason that how to support these individuals will differ as well, with some methods being more effective for some maths anxious individuals than others.

> Kirkland (2020) suggested that a "one-size-fits-all" approach to Maths Anxiety interventions will be ineffective, as the experience differs for each maths anxious individual.

Looking at Maths Anxiety as a "problem to be fixed" may be too simplistic. Perhaps, by understanding the individual's experience on a deeper level, working towards shifting their mindset and emotions towards Maths, and behaviour, to be more positive may work to lessen their anxious response to mathematical situations. Therefore, a better way to frame the question may be *"How can we understand Maths Anxiety better, and respond to this?"*

While it certainly isn't a magic wand that we can provide, this chapter will discuss various attempts to address Maths Anxiety that have been explored and reflect on their potential effectiveness, including:

- Targeting Maths skills and understanding
- Flipped learning approach
- Peer mentoring
- Cooperative learning
- Teacher impact
- Addressing avoidance
- Increasing self-awareness
- Breathing techniques
- Relaxation techniques

- Systematic desensitisation
- Expressive writing
- Psychodrama therapy
- Stories and storybooks
- Animal therapy
- Cognitive restructuring
- Reappraisal
- Mathematical mindset
- Mathematical Resilience

> Sheffield and Hunt (2007) identified that interventions to support maths anxious individuals should be personalised and focus on daily tools that the individual can implement successfully.

While these strategies will primarily be spoken about in the context of the classroom, it is important to highlight the positive impact that changes within the home environment can have as well. Therefore, when reading the different strategies within this section, reflect on not only the applicability within your own classroom setting, and the whole school, but also how these strategies might be utilised at home and other contexts outside of school. These can be communicated with parents through methods such as parent workshops, parent meetings, written communication, or even videos and online resources sharing top tips. However, if the strategies you want to embed are shared with parents, it is important that the same message and support to students are mirrored in the home environment, to support maths anxious students as best we possibly can. It is also essential to note that some strategies require specialist training; teachers or parents certainly shouldn't feel obliged or pressured to go beyond what they feel comfortable with.

Despite the fact that the academic literature on Maths Anxiety has come on leaps and bounds over the last few decades, empirical research into targeted strategies for addressing it is still quite limited. Nevertheless, when we consider what evidence there is, alongside anecdotal accounts, the picture looks positive. Since the 1990s, an upsurge in interest in Maths Anxiety has meant that interventions have taken even more twists and turns, trying out strategies that are based on developments in our understanding of the phenomenon. In particular, compared to the pre-1990s, the rise of digital technology has provided ever-increasing opportunities to improve Maths education, including the potential for addressing psychological barriers such as Maths Anxiety.

Looking at meta-analyses of tried and tested interventions can be really useful, as this provides an overview of existing methods and their effectiveness. This can highlight what is available for us to support maths anxious individuals, and what may or not work for any one individual.

> Hembree (1990) conducted a meta-analysis on the effectiveness of interventions for Maths Anxiety, which is still often referred to. Here are some interesting points:
>
> - Whole-class approaches to reducing Maths Anxiety were, on the whole, ineffective. This included changes to the mathematical content that was taught, in addition to the use of various types of equipment such as "microcomputers" and calculators. It also included different ways of presenting material, including class size and the pace of learning.
> - Whole-class psychological treatments for Maths Anxiety were also reported as ineffective.
> - On the other hand, interventions that took place outside the classroom showed a lot more promise.
> - Relaxation strategies had a positive effect, but not a statistically significant one.
> - Strategies using systematic desensitisation did demonstrate a significant reduction in Maths Anxiety.
> - Interventions based on cognitive restructuring were also successful in reducing Maths Anxiety.
> - A combination of systematic desensitisation with cognitive restructuring or relaxation training produced the greatest reduction in Maths Anxiety, although this combined approach was only marginally better than the effect of systematic desensitisation alone.

In this chapter, we mostly focus on specific approaches to address Maths Anxiety, underpinned by empirical research findings. This is useful in terms of unpacking those approaches to understand the relative impact of them. While there are several studies that have taken a highly targeted approach regarding interventions of a specific nature, it should be acknowledged that various studies have combined strategies into a single intervention. As we will go on to discuss, this is likely to be the best way forward if we want to address Maths Anxiety, taking into account the wide range of individual differences and factors associated with it.

> Passolunghi, De Vita, and Pellizzoni (2020) devised an intervention for fourth graders that involved a multistage process, including (1) knowledge and recognition of emotions through playful activities; (2) understanding what is causing changes in emotion, and what changes are observed in the body associated with changing emotions; (3) a focus on stories about other's feeling towards Maths; (4) breathing exercises and safe place visualisation; and (5) turning negative thoughts into positive ones. The intervention led to a significant decrease (with a large effect size) in Maths Anxiety compared to a control group.

Before discussing different strategies to support maths anxious students in detail though, it is important to highlight the barriers we currently face when utilising existing research to support students in our own classrooms.

While meta-analyses are powerful, they can also highlight areas of caution. They don't often provide much in the way of detail concerning the intervention strategies reported in individual studies. However, looking closely at the relevant individual published papers also reveals that they too are rather limited concerning the level of detail, particularly the procedural detail, to fully understand the strategy that was tested. This usually means that interested people are left to contact the researchers concerned to personally request the information that is needed if they want to try the strategy with their own students. In some ways, though, this isn't always a bad thing. A reliance on replication or copying an existing strategy can sometimes mean the essence of what that strategy is based on is lost; that the procedural detail overtakes the general underlying principles of the strategy.

An even closer look at more recent published literature on Maths Anxiety interventions also shows us that there are a limited number of attempts to test specific types of strategy. This can be troublesome for several reasons, including the lack of reliability testing, which tests the extent to which the same findings are reproduced on multiple occasions. This is especially problematic when a strategy is based on a general principle but different studies vary the strategy slightly, meaning there are very few attempts to re-test the effectiveness of exactly the same strategy over time or with different groups of students. Based on this, while it is exciting when empirical findings on a successful new strategy are published, we should take care not to get carried away and assume the strategy is the go-to solution for Maths Anxiety.

We should also note there are several things to consider when it comes to ways to address Maths Anxiety. There is published academic literature, often based on empirical attempts to test strategies that have been implemented with a specific group of students in a specific educational context. These attempts are usually very systematic, whereby researchers have tried to control for variables they think could impact a student's Maths performance or Maths Anxiety. Such control enables us to have greater confidence in the effect of the chosen Maths Anxiety strategy, whether that's a negative effect, no effect, or a positive effect. However, these kinds of studies, which are often experimental based, can sometimes mean quite small groups of students are tested and, as mentioned earlier, we don't always know whether the same effects are likely to occur with a different group of students.

Either way, teachers should always think about their own students and the myriad of ways in which those students might be different – both in terms of variation *within* the group but also differences *between* that group and other groups of students. For example, a teacher will get to know a student's personality, their motivations, attitudes, and behaviours associated with Maths learning and testing – and how these may have changed over time. These features of an individual student all need to be taken into account when considering which strategies to adopt, the order in which they should be implemented, and the exact ways in which selected strategies are embedded. Simple, but important school- or class-based factors will also need to be considered from a pragmatic perspective, such as the age of students, the size of a class, or the resources available. Factors such as language, socio-economic status, and attitudes towards learning more generally might all need to be acknowledged when asking the question of whether a strategy that was successful in one school will be successful in another.

Other factors might need to be considered too, including the likely resources involved in a particular intervention, such as time, number of people required for implementation, any training required, and any physical materials that might be needed. Based on this, to some extent, teachers should form a judgement that takes into account what researchers report, their own interpretation of published findings, the resources required, the specific needs of their students, and the relative benefits derived from interventions when compared against others that are available.

> Sammallahti et al.'s (2023) meta-analysis found that interventions involving students over 12 years old showed the biggest decrease in Maths Anxiety, highlighting the possibility that there are more challenges when it comes to addressing Maths Anxiety in younger children.

Aside from empirical attempts to design and test targeted interventions for addressing Maths Anxiety, teachers can also take note of less structured, more informal attempts, but with the same objective. These might be recorded by teachers in formats other than peer-reviewed academic journals, such as through blogs, or magazine and newsletter articles. This more anecdotal approach may not provide the same level of statistical detail regarding the efficacy of a strategy that has been tried or tested, but the teacher concerned might give particular insight based on their extensive experience in the field, which is always worth paying attention to.

Published Reviews of Interventions

We recommend that you explore the following academic reviews of empirically tested Maths Anxiety interventions:

Hembree (1990) - a meta-analysis
Petronzi, Hunt, and Sheffield (2021) - a narrative review
Balt, Bornert-Ringleb, and Orbach (2022) - a systematic review
Codding et al. (2023) - a meta-analysis
Sammallahti et al. (2023) - a meta-analysis

Decisions around the selection of Maths Anxiety strategies will need to further consider the cultural context in which the teacher is working, acknowledging sensitivities when dealing with students' emotions, but also the more general school policies. Based on all these considerations, we advise teachers to acknowledge the range of strategies that have been proposed, tested, and discussed and to make informed decisions about the suitability of certain strategies with their own students and learning context in mind.

A Note on Effect Size

While we have purposely avoided providing too many statistical details when describing research findings, it is worth taking a moment to mention the importance of which information is used to make a decision about the effectiveness of interventions, including what is meant by the term "effect size" for those who are unfamiliar with it. Researchers might report that an intervention resulted in the desired or expected effect. This is often based on a comparison of average scores on a particular measure, representing the variable of interest, e.g. Maths Anxiety. The comparison would usually include two or more groups, such as those who experienced an intervention and those who did not, but also scores before an intervention and those after it, to observe any change.

A common approach by researchers is to run a statistical test, which is based on probability testing. Typical tests would be analysis of variance (ANOVA) or t-tests (or their non-parametric equivalents), resulting in the reporting of a p-value. In the behavioural sciences, a p-value of .05 or less is usually regarded as acceptable in order to accept the corresponding research hypothesis or reject it if the value is above .05.

As such, the first thing to consider is whether researchers are making conclusions about the effectiveness of an intervention based on descriptive statistics alone, or whether they are analysing the data more robustly, using statistical probability testing.

However, there are issues with a reliance on probability testing, so it is always worth looking beyond that approach in order to make your own judgement about the results of a particular intervention study that are reported. One of the main issues is sample size (the number of research participants involved in a study). A study with a larger sample size increases the likelihood of a probability value being lower, and closer to the accepted value of $p = .05$. In some cases, interventions might involve impressively large numbers of students, resulting in a p-value of less than .001, for example, meaning the result is highly statistically significant. This, however, might say more about the number of people involved, rather than the actual impact of the intervention. In other words, if researchers compared changes in Maths Anxiety between a group of students receiving some form of intervention (the experimental group) and those who did not (the control group), the change in Maths Anxiety between the groups, over time, might be statistically significant but quite small in real terms.

To present the actual size of the effect of the intervention, researchers will often report an "effect size" statistic. This is usually a standardised value whereby the smaller the number, the smaller the effect. Likewise, a larger effect size represents a larger effect, or difference between means. In intervention studies within education, two effect size statistics that are commonly reported include Cohen's d and Pearson's r. It is beyond the scope of this book to delve into these too deeply, but it is useful to be aware of why these effect size statistics are reported, and how to interpret them.

As a rough guide, values of d of 0.2, 0.5, and 0.8 represent a small, medium, and large effect, respectively. In that case, anything below 0.2 (getting closer to zero) is particularly small. The higher the value is above 0.8 (it can go beyond 1), the bigger the effect.

> Regarding r, values of 0.1, 0.3, and 0.5 represent a small, medium, and large effect, respectively. Unlike d, r cannot exceed a value of 1.
>
> If we consider studies where the p-value and the effect size do not quite match, especially where there is a statistically significant result but a small effect size, this poses a bit of a dilemma for the person trying to interpret the results; was the intervention successful? Should teachers be adopting the intervention in their classroom? It should be noted that even small effects, such as minimal reductions in Maths Anxiety, could still be of educational and therapeutic importance. As such, it is not always helpful to rely too heavily on any one statistical approach to testing an intervention's effectiveness.

With all this in mind, let's explore some attempts by researchers to empirically test various strategies to tackle Maths Anxiety.

Targeting Maths Skills and Understanding

When looking at mathematical attainment within the context of Maths Anxiety, we have seen that it can be an important factor. Typically, a maths anxious individual is more likely to find Maths difficult or at least more challenging, whether in general, or in relation to certain mathematical concepts. Therefore, it stands to reason that if an intervention focuses on improving mathematical understanding and building specific Maths skills, then their experience of Maths Anxiety may also improve.

Research has shown that by focusing on improving an individual's mathematical understanding, their attitudes and affect towards Maths improve. This has a positive implication regarding Maths Anxiety, yet the majority of studies that focus on the impact of improving attainment focus on affect more generally, rather than specifically on Maths Anxiety. Even so, this provides us with a good place to start – by improving a student's mathematical attainment, underpinned by an improvement in Maths skills and understanding, their levels of Maths Anxiety may be reduced.

> Supekar et al. (2015) created an intensive one-on-one cognitive tutoring programme with 46 children aged between eight and nine years old, that lasted for 8 weeks, designed to improve mathematical attainment and reduce Maths Anxiety. Their programme was adapted from Maths Wise and combined conceptual instruction with speed retrieval of number facts in over 22 lessons which increased in difficulty. They found that the students who were highly maths anxious showed a significant reduction in their Maths Anxiety levels after they had completed the tuition programme. They also measured neurobiological mechanisms associated with Maths Anxiety, such as the basolateral amygdala, and found that those who had the biggest reductions in their levels of Maths Anxiety also had the largest decreases in their amygdala reactivity. This highlights the important role that the amygdala, associated with the emotion of fear, has in the experience of Maths Anxiety, and that a focus on using tuition to improve mathematical attainment can have a positive effect on Maths Anxiety.

As teachers, this may be a relatively uncomplicated intervention to put in place to support maths anxious students. Academic interventions relating to mathematical attainment may already be occurring within your schools, so working out the possibility and logistics of conducting these interventions may be relatively straightforward. They will also be cheap to run, as they will probably utilise existing members of staff, yet consideration will need to go to the expense of time. Existing studies that focus on improving mathematical attainment and Maths Anxiety are usually longer term than other interventions available, so require a lot of forethought, planning, resourcing, and time. How much time can you afford to give to the intervention, alongside teaching the rest of the curriculum, and other children? Three times a week, for 45 minutes each time, seems unrealistic, but would once a week have the same positive effect? Or would three times a week for 15 minutes still have a positive effect, due to the regular, short intervals between sessions? A further consideration is the age of the students participating, as a 45 minute session for children in Key Stage 1 will be challenging in terms of them remaining focused and on task, whereas 15 minute sessions would be much more suitable.

Chelsea's Experience

To address Chelsea's Maths Anxiety, she had one-on-one tuition which focused on gaps in her learning, primarily related to the manipulation of number. This happened twice a week for 15 minute sessions over the period of a half-term; once during assembly time and once during form time. Focusing on concepts that Chelsea found tricky had the potential to evoke anxiety. However, reducing these down to a level that she felt comfortable with, before increasing the difficulty as her level of comfort and understanding improved, appeared successful. Chelsea shared that she felt more confident in using strategies such as written column subtraction and knew how and when to use it. Chelsea's teacher felt that her increased confidence in manipulation of number also reduced the anxiety that she had previously experienced when approaching new Maths problems.

A lot of consideration needs to occur here though, in order to successfully implement a Maths Anxiety intervention based on improving mathematical skills and understanding. Knowing your students well, including how they feel they learn best, what areas of Maths they would like to improve upon, and what they need to improve upon, is key to the success of these interventions. Heidi has conducted many one-on-one intervention sessions with maths anxious students within primary schools, with varying outcomes. From reflection, those with the most positive impact on Maths skills and understanding in addition to Maths Anxiety have been conducted where there has been a strong understanding of the student, including their needs, their interests, and their strengths and difficulties in relation to learning Maths. Interventions that had the most success over the years were with students where there was a good relationship built on trust, which had developed over a period of time, compared to sessions with students who were less known, and where there wasn't an existing relationship. Therefore, it is essential to know maths anxious students well before implementing any intervention, in order to work towards the best outcome for them.

Trust within a group intervention may also affect the outcome and success, compared to a one-on-one intervention. While it may not be feasible to conduct one-on-one interventions due to the resources and time required, we should be mindful that a group dynamic may also change the engagement levels and openness of students. Their wants, interests, and needs may also not align, making it difficult to ensure that the interventions are personalised and tailored to each maths anxious student.

It is possible that the heightened teacher attention afforded by one-to-one or small group sessions might be what students need to address underlying deficits in Maths skills and understanding that have ultimately influenced feelings of Maths Anxiety. But there will be individual variance within this. Some maths anxious students will find it extremely difficult within this setting to manage their emotions due to perceived pressure and potentially high expectations felt from working one-on-one with a teacher. Within a small group, they may also be very mindful and anxious about getting something incorrect in front of others. Therefore, it is important to again reflect on the individual student and how they may react within certain situations, to ensure that this intervention is the most suitable for them.

This approach to improving Maths Anxiety may utilise home learning as consolidation of the session contents, or providing videos to support the concepts learnt. Introducing new content at home though, without a teacher present, may not support this particular strategy, unless it is part of the larger pedagogical approach known as "flipped learning," which will be discussed next in this chapter. Ensuring that parents are informed about the focus on attainment to address Maths Anxiety may allow home learning to be more effective and provide an increased sense of support for the student from both home and school. It is important to highlight to parents though, that additional work is not a requirement, so it's key to educate parents on what you are doing, why you are doing it, and how they can support this within the parameters set out by you, the teacher.

> Willis (2010) created an overview of strategies designed to improve students' attitudes towards Maths and lead to improved mathematical performance. Suggestions consist of:
>
> - Ensuring that the goals set in place are achievable, and meaningful, for each student;
> - Utilising students' strengths and interests;
> - Increasing the perceived utility of Maths;
> - Discussing misconceptions and how to respond to making mistakes.

From a deficit perspective, while aiming to improve Maths skills is useful as a way of addressing Maths Anxiety, it may not be the most effective approach to reduce feelings of anxiety when considered against other available strategies. This supports arguments that suggest Maths Anxiety is not simply a result of poor Maths skills; it is much more complex than this.

It is suggested that teachers should not consider a "one or the other" approach to tackling Maths Anxiety in their students though, as the development of Maths skills and therapeutic approaches both appear to be helpful in addressing the issue. Findings from research emphasise the importance of separating the anxiety from mathematical skills and understanding; these can both be hugely important aspects of a Maths student's journey, and the needs of the individual should be identified and acted upon accordingly. Having the knowledge and tools to achieve this is therefore vital.

> Codding et al. (2023) compared interventions for Maths Anxiety and noted the relative importance of therapeutic approaches to tackling Maths Anxiety compared to non-therapeutic approaches, which focused on targeting mathematical skills. Their meta-analysis focused on 17 studies which used school-based interventions for secondary school students. Compared to skill-based interventions, they found that therapeutic interventions had a greater effect on reducing Maths Anxiety. Conversely, the interventions that targeted Maths skills led to a greater improvement in Maths performance. This suggests that non-Maths-based interventions, specifically therapeutic ones, continue to be successful in reducing Maths Anxiety.

Flipped Learning Approach

Using the flipped learning approach means that what used to be classwork, the traditional teaching part, is done at home through videos created by the teacher, and what used to be homework, such as assigned problems, is completed in class in a more dynamic and interactive way. In this manner, students are introduced to the learning material before the lesson, so that they can fully utilise the lesson to deepen their understanding through facilitated and collaborative learning, as appropriate.

> Flipped learning has been described as "an educational technique that consists of two parts: interactive group learning activities inside the classroom, and direct computer-based individual instruction outside the classroom."
> **–Bishop and Verleger (2013, p. 5)**

In Maths, flipped learning could happen with the use of online learning materials, where the students are introduced to Pythagoras Theorem, for example, for their home learning. Then, the subsequent lesson in the classroom would focus on strengthening this understanding through associated tasks, and developing students' reasoning and problem-solving skills based on their conceptual understanding of Pythagoras Theorem. In this manner, the teacher acts as a facilitator of learning, or a coach, instead of using the lesson time to teach knowledge. This maximises time for student learning and has been found to motivate students and lead to more effective learning and outcomes.

Considering what we know about the role of evaluation in the context of Maths Anxiety, and based on the apparent benefits of peer-learning approaches to Maths, it is perhaps no wonder that researchers have found a reduction in Maths Anxiety through utilising a flipped learning approach. From an anxiety perspective, the flipped approach would mean the student would not be contending with a fear of evaluation while simultaneously engaging in instructional material, but they would also benefit from the support of others while engaging in actual problem-solving.

> Niaei, Imanzadeh, and Vahedi (2021) researched the effect of flipped learning upon Maths Anxiety and Maths performance of 56 fifth-grade students in Iran. They found that Maths Anxiety decreased significantly within the flipped learning class, compared to students within the traditional learning class. They suggested that this was due to flipped learning facilitating deeper learning, increasing motivation, and allowing students to have higher levels of involvement and set their own goals.

Importantly though, to embed flipped learning as a pedagogical approach to reduce Maths Anxiety, it needs to be acknowledged that not all students have access to the internet, or a computer, outside of the classroom. This digital divide became widely apparent during the 2020 lockdowns due to COVID-19, leading to learning gaps due to the lack of resources available to all students. This pedagogical approach also requires teachers to rely on students. If the students don't complete the initial learning at home prior to the lesson, they cannot engage with the tasks within the classroom. This emphasis on student agency limits the applicability of using this approach to reduce Maths Anxiety in younger students, who may find this difficult. While teachers can inform parents and involve them within the home learning aspect of flipped learning, the focus, in order to be truly effective, should be on student agency and independence. Therefore, using the flipped learning approach may only be applicable later on in secondary schools and further education, when students often become more accountable for their learning.

> Segumpan and Tan (2018) used a flipped classroom approach to successfully reduce Maths Anxiety in secondary school students in the Philippines. Based on their findings, the researchers recommend that teachers should avoid using home-based activities such as assignments as a final summary of the previous lesson. Rather, they should use them as an introduction and pre-discussion of the next lesson. Their findings suggest that this alternative way of learning has the benefit of reducing Maths Anxiety.

In addition to this, a flipped learning approach may require a lot of pre-planning from the teacher, gauging what students need to know prior to the facilitation lesson within the classroom, and ensuring that the online learning done independently outside of the classroom is accessible by students of varying mathematical abilities.

Interestingly, flipped learning with teacher-created videos has also been reported to be effective in reducing Maths Anxiety in trainee (pre-service) primary teachers as well. This leads us to ask another important question: can teachers be maths anxious, too? We will discuss this in Chapter 7.

Peer Mentoring

Learning often takes place in the context of others, as students are usually surrounded by their peers. It therefore makes sense to consider ways in which peers might provide a supportive role in helping those who struggle with Maths Anxiety, through peer mentoring. This strategy is based on one student providing support to another student who is of the same age, or the same stage within their education, but has the skills and previous experience to allow the individual they are mentoring to progress and learn from different perspectives. Peer mentoring may happen in Maths classrooms on a casual basis, such as working in a mixed ability pairing, or it may be an organised intervention that happens inside and outside of the classroom, with dedicated time for the mentoring process to occur.

Peer mentoring has been found to benefit both the mentor and the mentee. The mentor is able to develop their confidence and skill set, including the skills of communication, leadership, and active listening. The mentee experiences the benefit of feeling supported, and being provided with an opportunity to process and regulate their emotions and negative beliefs. As an effective tool for mental health and well-being in schools, peer mentoring has also been found to reduce Maths Anxiety in students and have a positive effect on Maths performance. That said, empirical attempts to design and systematically test interventions to improve Maths Anxiety based on this concept are limited, so it is difficult to generalise the effectiveness of peer mentoring in this way. Nevertheless, positive impacts of peer mentoring have been found on levels of Maths Anxiety, highlighting this as a potential strategy to support maths anxious students in the classroom.

> Cropp (2017) used a peer mentoring approach with a handful of maths anxious secondary school students. This involved student mentors providing encouragement and demonstrating skills to cope with getting "stuck" with Maths. Pairings then worked together for four one-hour sessions over a period of six weeks. A positive effect of the intervention was reported, stating that all students had a positive attitude towards the intervention and that most reported a reduction in Maths Anxiety. It is difficult to generalise from the small sample size, but the results were promising and suggest peer mentoring is worth exploring further as a method for addressing Maths Anxiety in the classroom.

Therefore, peer mentoring could be a useful tool for teachers to utilise within the classroom, whether this is through mixed-ability pairs to allow peer mentoring to occur within the lesson itself, or creating certain points within the lesson for students to stop and reflect, and speak with their peer mentor. An alternative way to utilise peer mentoring may be outside of the classroom, even with a student from a different Maths class. This may improve the level of openness between mentor and mentee, as this might limit the associated pressure of the mentor being in the mentee's class.

> Moliner and Alegre (2020) researched the effects of reciprocal peer mentoring on Maths Anxiety in 420 secondary school students in Spain. Statistically significant reductions in levels of Maths Anxiety were found in those who participated in the experimental group, using peer mentoring, while no changes were found within the control group. A positive effect of peer mentoring upon Maths Anxiety was found regardless of age or gender.

However, you decide peer mentoring would work best in your setting, it would require training sessions for mentors, including outlining what is expected, how best to support mentees, and what to do if they have any concerns or worries. This strategy also requires engagement from maths anxious students and for them to feel that they are able to be open and honest, while trusting their mentor. While this can take some time to develop, the suggested outcomes are likely to benefit the maths anxious individual.

Raj's Experience

Raj's teacher recognised that Raj's confidence in Maths lessons was dropping lesson on lesson. Subsequently, Raj was paired with a student who repeatedly demonstrated high confidence when learning Maths. At first, Raj felt a little intimidated and didn't enjoy pair tasks, but his mentor displayed patience, compassion, and understanding. Raj gradually felt able to make more genuine attempts at the Maths problems he was presented with, realising that his mentor made more errors than he had initially assumed. Raj noticed that his mentor seemed to learn from these errors, which spurred Raj on. While this took some time to have an effect (approximately a half term), the positive impact on how Raj engaged with Maths and viewed mistakes, as well as his self-belief, was clear.

Cooperative Learning

Levels of Maths Anxiety have been reduced through using cooperative learning as a pedagogical approach within the classroom. Cooperative learning, sometimes also referred to as group learning or collaborative learning, requires students to work in groups to achieve the outcome of the lesson. Cooperative learning is more specific than this though, as it structures the roles within the group and is overseen by teachers in the role of facilitator, or coach. Working within a group on Maths tasks can provide an opportunity for interdependence and interaction that means students do not feel solely responsible for finding solutions to Maths problems. This also allows them the chance to discuss their thoughts with their peers.

> Mehdizadeh, Nojabee and Asgari (2013) found that cooperative learning led students to be significantly less maths anxious, more likely to seek help, and be less avoidant when solving Maths problems.

An example of cooperative learning within a Maths lesson may be where students are asked to work together to find different shapes which have the same area as each other. One student might have the role of delegation, while another might be an innovator, an encourager, the checker, or the reporter. Making sure the students know what each role entails, and how to do this successfully, can lead to positive outcomes for all students, and reduce feelings of anxiety. Perhaps, it is the smaller role within a larger task that is less anxiety provoking, or that they are part of a group with shared responsibility, that reduces the anxiety felt.

It is also important to note that this strategy does require each student to effectively participate within the group, but this is more likely to happen with assigned roles, rather than working in a group with no defined responsibilities. Therefore, it may be seen that cooperative learning limits avoidant behaviour and improves engagement. This may not be the case for every maths anxious individual though, as having a role that they deem to be stressful, such as being the group's checker and being responsible for accurate calculations, can exacerbate their feelings of anxiety. Therefore, the teacher must ensure that the role is suited to the maths anxious student and regularly check throughout the lesson that their role is not evoking anxiety at any point.

> Pantino and Hondrade-Pantino (2021) used a series of creative writing tasks (vocabulary paragraphs, numbers stories, poems, advice columns, and research activity) based on Maths to see whether students' engagement in the tasks reduced their Maths Anxiety. They found that Maths Anxiety, specifically Maths Test Anxiety, was reduced after the tasks, but only in the condition in which students worked on the activities as a group. The students who worked on the tasks individually showed no reduction in anxiety levels, nor did the group of students in the no-writing condition. The extent to which the writing tasks focused on the emotional or worry aspects of students' thinking is unclear, but the tasks appeared to be a useful way of encouraging cooperative learning as a method for reducing Maths Anxiety.

Perhaps, one reason for the success of cooperative learning in reducing feelings of Maths Anxiety is that students feel more able to ask others for help if they are finding Maths difficult. The exact reasons for this are unclear, but a reasonable suggestion is that feelings of shame and fear of failure or judgement are addressed through cooperative learning. Research has shown that students do not form new peer networks based on their levels of Maths Anxiety, so teachers can play a vital role in facilitating interactions between students with different levels of Maths Anxiety.

> Lavasani and Khandan (2011) explored the role of cooperative learning in reducing secondary school girls' Maths Anxiety. Unfortunately, they do not provide much procedural detail concerning the nature of the cooperative learning approach, but they did specify that the approach helps students to tackle their learning problems through discussion, consultation and help-seeking, and that it provides alternative solutions for solving problems and learning the various problem-solving strategies with the help of their peers and teacher. A comparison of this approach against a traditional learning and teaching approach resulted in a reduction in students' Maths Anxiety. They also found that cooperative learning reduced avoidant behaviours and increased help-seeking behaviours.

Using cooperative learning within the Maths classroom may require an increased level of forward thinking from the teacher, and time spent explaining expectations and each different role required within the group. However, once students are aware of these, the time spent on this in future lessons is likely to be reduced, as cooperative learning becomes a part of the classroom culture.

Teacher Impact

We know that teachers are influential when it comes to students' Maths Anxiety, both in creating feelings of anxiety, but also having the ability to improve students' experiences as well. A typical behaviour that takes place in the Maths classroom is where a teacher verbally asks the class, or individual students, Maths questions. The anxiety that some individuals experience at the prospect of being asked a question in front of their peers should not be underestimated. Students will sometimes say they spend an entire Maths lesson in fear of being asked a question in front of others. Such preoccupation clearly takes a student's attention away from being able to concentrate on vital features of their lesson, which will ultimately impact the depth of their learning. So, how can teachers improve this for maths anxious students?

Is it best to take an inclusive approach by not leaving out individual students when asking questions, providing an opportunity for students to give a correct or appropriate answer and thus supporting confidence building? Or, is it better not to ask anxious students questions at all? A professional judgement has to be made regarding this, and what is helpful for individuals. Again, knowing the students well will support this decision making, and ensuring that the well-being of all students is at the forefront. One thing for sure is that it is better that a highly maths anxious student feels psychologically safe enough to learn in a Maths lesson, rather than spending the entire time only worrying about being asked a question by their teacher.

We should also bear in mind that a verbal question means that the student has to maintain that information within working memory, while also attempting to solve a problem, and also dealing with anxious feelings and thoughts. In other words, there is quite a lot for the maths anxious student to contend with, which often results in the common experience of one's mind going blank. Having questions written visually, where possible, will support maths anxious students within the classroom and provide them with a visual cue when engaging with Maths questions.

> **Callum's Experience**
>
> Unbeknownst to Callum's teacher, Callum spent most of his Maths lessons in fear of his teacher asking him a question in front of his peers. Through a conversation with Callum, his teacher recognised that his fear was taking over, leaving no mental space to concentrate. Callum's teacher struck a deal with him that she would not ask him any questions in front of others for the remainder of the term. A visible reduction in anxiety followed, and Callum and his teacher continued to talk about the effect that questioning was having on him. This helped Callum identify his feelings, and the likelihood of his "worst fear" coming true; being embarrassed in front of the class. To transition back to answering questions, Callum's teacher asked the class to write down their answers on whiteboards and show them to her. If Callum got it incorrect, the teacher wouldn't ask him related questions in front of the class but would support him individually within the lesson later on.

A potential middle ground following a verbal Maths question to the class is to ask all students to write their answer on a miniature whiteboard, consequently holding up their board for the teacher, and not the rest of the class, to see. This can alleviate some of the worries that students have, but it also has the added benefit that the teacher can see multiple responses, providing a broader view of the level of understanding of the whole class.

Offering students the chance to discuss their thoughts with a partner, or a small group, may also alleviate feelings of anxiety when asked a verbal Maths question. Through the use of talking partners, or "think, pair, share," as a couple of examples, giving maths anxious students the opportunity to discuss the question posed with their peers may provide them with feelings of security knowing that they got the question correct, or feeling secure that their peers have told them the answer. Therefore, if they are asked to answer aloud in front of the class by the teacher after this discussion, they may feel more confident in doing so. This strategy does require a deep understanding of the individual's experience of Maths Anxiety, however, as it could have the opposite effect. Having to discuss their thoughts in a small setting, with just one or two of their peers, can heighten an individual's feelings of anxiety if they feel that they have to participate and don't know how to answer the question. This further demonstrates how tackling Maths Anxiety cannot be a "one-size-fits-all" approach.

Teachers will often give a reassuring message when encouraging students to give an answer to a verbal Maths question, demonstrating a degree of empathy or belief in the student. While this is well intended and can be helpful in supporting a students' confidence, it can sometimes have the reverse effect in highly maths anxious individuals. Anything said aside from the question itself can draw their peers' attention to the student, thus exacerbating feelings of anxiety. Reassuring statements from the teacher affirming their belief in the student's ability can also impact their sense of expectations: *"the teacher is confident I know the answer so I'm going to look even more stupid if the answer is wrong."* Again, this can increase anxiety so it is important to reflect on the impact a teacher can have on maths anxious individuals with their words and actions.

> Núñez-Peña, Bono, and Suérez-Pellicioni (2015) suggested that giving students feedback about previous errors made in assignments reduces the impact of Maths Anxiety on students' Maths performance in tests.

Moreover, having greater awareness of the subtle effects of things like perceived rewards and punishment might be a good first step in minimising negative associations with Maths at a young age. Using teacher development techniques such as peer teaching, lesson coaching, informal lesson observations, or keeping a reflection diary about certain uses of languages and behaviour can help teachers develop the awareness of the impact they are having, and work towards resolving this in order to support maths anxious students. Increasing the awareness that teachers have of the power of their language choice and behaviours, and the impact that this can have upon students, is an excellent way in which to not only better understand the experience of Maths Anxiety, but to provide an environment which works to support students through providing a safe, and non-judgemental, space.

Another way in which teachers can have a positive impact on Maths Anxiety is to reflect on the tasks used within lessons, and how to support the different needs of students. As we are aware, the role of the working memory is key within the experience of Maths Anxiety. Therefore, by reflecting on whether the tasks given are too taxing on working memory, and changing them in order to reduce this load, will be beneficial. For example, if a task is based on multi-step word problems and discussing these with a partner, a maths anxious individual needs to recall prior learning, potentially hold numbers in their head while solving multiple steps of a problem, and listen to their partner's thought processes as well, in what is likely to be a classroom full of students talking. Instead, providing visual aids such as multiplication grids, and reminders of how to use certain methods, whether these are placed on the table or on a classroom display, can help reduce the pressure on students' working memory. On top of this, encouraging each step to be written down on a whiteboard can alleviate the need to hold any numbers or calculations in their mind and prevent the need for mental maths when it's not a necessity.

Manipulatives within lessons can also be beneficial, such as using dienes, counters, or other physical resources to help represent or manipulate numbers, as this is again providing a visual and physical way in which to process the task at hand. For example, instead of holding the number 42 in their head, a student may use base 10 apparatus to create this number and then focus on the next number they need to create. This again highlights the importance of knowing the students, including their abilities, needs, and preferences, in order to ensure that they are adequately supported within lessons, and to have a positive impact their anxiety. Sharing this knowledge with parents will again be useful for how they may best support their child at home when learning and using Maths. However, research into the use of manipulatives for maths anxious students is limited, so it will be interesting to observe developments in this area.

Addressing Avoidance

In Chapter 4, we discussed how Maths Anxiety is associated with avoidant behaviours, often representing short-term coping strategies to avoid Maths. This could be at the micro level, such as speeding through a Maths problem at the expense of accuracy, or even not responding to a Maths question at all. It could also be at the macro level, such as deciding not to study certain courses because of the real or perceived Maths content, such as Maths-related GCSE subjects or programmes in further or higher education. It could also be a wide range of avoidant behaviours in between, including putting off doing Maths homework. Therefore, in tackling Maths Anxiety, there needs to be a focus on dealing with avoidant behaviours, starting with the lower level behaviours that are perhaps more easily addressed.

To do this, there needs to be acknowledgement that avoidant behaviours are taking place, and while individuals might be consciously aware of wanting to avoid Maths at times, they do not always associate some of their behaviours with such avoidance or acknowledge that using avoidance strategies will not help them in the long run.

Therefore, one approach could be to discuss with individuals whether they do engage in avoidant behaviours and, if so, which ones. The next step, in partnership with the individual, would be to consider what rewards could be put in place when avoidance is, well, avoided. It could be a reward for not completing Maths homework at the last minute for example, or a reward for genuine attempts at solving Maths problems.

This type of open dialogue may be better suited for older students, as they may have more awareness of their actions and the associated consequences. However, conversations such as this are still suitable to use with primary school children, but with language that is tailored more to their age and understanding.

Myra's Experience

Myra's teacher noticed that Myra would sometimes appear stressed when stuck on certain Maths problems. Myra's response would be to write a nonsensical answer and quickly move on, often meaning that she would experience the same thing on subsequent Maths questions. Myra's teacher decided to start rewarding students for seeking help if they were troubled by particular Maths problems and further rewarded students if they made genuine attempts after initially struggling. This led to a noticeable attempt from Myra to ask for help, although only in instances when her peers would be less aware of the support.

The concept of genuine attempts is important in the context of Maths Anxiety, but it is not always easy for teachers to identify whether an attempt, either written or verbal, is indeed genuine. This is where some judgement, perhaps based on existing knowledge of the student, is required. This approach of rewarding non-avoidance would need to be carried out in conjunction with other strategies. After all, simply putting a highly maths anxious student in more and more Maths situations is not suitable on its own. The student would need to feel prepared in recognising and managing their anxiety once in those situations.

Increasing Self-Awareness

Many researchers and teachers use self-report Maths Anxiety scales to assess the degree of Maths Anxiety, both at an individual and group level. As discussed earlier in Chapter 3, questionnaires of this nature have a lot of value. However, one understated benefit of these self-report scales is the way in which individuals might use them themselves as a method of self-reflection.

It is important to bear in mind that researchers, teachers, and various other professionals have a certain amount of awareness of Maths Anxiety as a construct. They are aware of its existence and the ways in which it develops or manifests in different contexts. Many students however, particularly younger students, might not have ever fully reflected on their own feelings towards Maths. They might be aware of their general attitude towards Maths, but they might not have deconstructed those feelings or attitudes in a way that separates out anxious feelings from, say, feelings of confusion or boredom. Similarly, they might never have properly considered the way in which they feel across a range of specific contexts.

Based on this, self-report scales offer a way of initiating thinking and discussion surrounding Maths Anxiety. They provide a tool that brings together the student, teacher, and/or other students to discuss feelings towards Maths in a way that might not have happened previously. In this context, students and teachers might even choose to ignore the total scores on such scales if the purpose is to discuss the ways in which students themselves relate to the specific items or situations listed. The process of reflecting on anxious feelings in different mathematical contexts can provide students and teachers with a greater understanding of where difficulties lie, including broad contexts such as home, school, or work. Of course, this depends on which scale is selected, which is why it is important to choose wisely.

Teachers might choose to use scales for this purpose in a variety of ways, including the facilitation of one-to-one conversations with students, or to allow groups of students time and freedom to discuss and express how they feel, whether that is followed by a conversation with the teacher or not. Depending on the individuals, or the class, and their age and ability to focus, it may also be an option to break down the questionnaire into chunks and use individual questions as talking points throughout the week. For example, beginning a Maths lesson with one question from a Maths Anxiety questionnaire may spark individual reflection and insightful group discussions. It is important to reflect on how these are used though, as to not elicit further feelings of anxiety or pressure from group or class discussions. Some maths anxious individuals may respond more positively to this approach within a one-on-one setting.

Beyond the use of self-report measures as a reflective tool, the questionnaire scores can also provide useful information ahead of choosing or implementing particular strategies for addressing Maths Anxiety. Teachers or other educational practitioners might use self-report measures as a way of simply determining which students are most in need of support. This should involve bearing in mind the earlier discussion concerning challenges with this, such as the absence of agreed requirements for what score might be deemed "high," and how a student's score might be deemed "low" or "high" relative to the scores of their peers. The decision might, at least partly, be a pragmatic one. Resources, such as time available, might dictate that only a specific number of students receive additional support for their Maths Anxiety. Therefore, at least in the first instance, it could be that a predetermined number of students in a class or cohort receive targeted support based on who scores the highest on the scale that is used.

Whether it is at the individual or group level, observing scores on particular subscales can help determine the exact content of Maths Anxiety interventions. For instance, a high score on a subscale associated with Maths testing or evaluation might give a focal point for understanding a student's thoughts about Maths testing. Such understanding is helpful if the aim is to then address those thoughts directly. Similarly, a high score on a subscale related to anxiety in Maths learning contexts could highlight a need to focus on – you guessed it – the context of learning Maths.

Beyond the total, or subscale, scores of self-report measures of Maths Anxiety, it is always worth paying attention to how a student scores on specific items. This can give a more fine-grained understanding of specific contexts that a student associates with anxiety, especially if they are not so forthcoming in identifying it or verbally describing it. Again, this can give rise to some useful discussion, along with highly targeted, context-specific strategies.

While self-report measures can kick-start a degree of self-reflection among maths anxious individuals, it is essential that these individuals are able to recognise feelings of anxiety as being distinct from other, related feelings. As we have already discussed, Maths Anxiety is related to a wide range of other thoughts and feelings, so it is likely to be helpful for students to better understand themselves – to understand what anxiety is and to disentangle their feelings of anxiety from anything else they might be feeling. This could be particularly important in younger students, and those in families and social groups in which discussion of feelings and emotions is less common. Improving self-awareness can lead to heightened action and positive effects upon Maths Anxiety, so it's important to embed the teaching of awareness within the classroom, so that not only are emotional reactions such as anxiety freely spoken about, but they are better understood in relation to how they can affect the person psychologically, as well as physically.

It is common for individuals, including adults, to have a basic lack of understanding when it comes to the physiological changes associated with anxiety. This could mean that a highly maths anxious student experiences stomach aches or headaches when they are in a Maths situation, or even in anticipation of being in one. Yet, they might not necessarily attribute such physiological experiences with feelings of emotional upset or worry. Younger children especially might ask their teacher if they can go to the toilet during a Maths lesson; this could represent a genuine increased need to go to the toilet, even if the student doesn't associate it with their negative thoughts and feelings. A specific instance such as this can be challenging for teachers, knowing when a student's need for the toilet is genuine or an escape route from the Maths lesson.

Increased self-awareness of bodily sensations as a function of anxiety can provide students with a greater sense of control. There are situations that result in peaks in physiological arousal in highly maths anxious individuals, typically evaluative, or potentially evaluative, situations. The panic that an individual experiences in such situations can be overwhelming, but more so if there isn't a fundamental awareness of why they are responding in a particular way.

Teaching individuals the fundamentals of the fight or flight response, and how this specifically relates to feelings associated with Maths Anxiety, can be surprisingly helpful for them. Simply increasing awareness of the reasons behind the changes in how they feel physically can minimise panic. Therefore, one approach is to include some basic teaching around the way in which our nervous system responds to a perceived threat. There are plenty of other sources available that provide useful materials on this, but the key message is that a perceived threat initiates quite an impressive chain of events in our brain and body, providing us with the means to deal with that threat. This includes mechanisms for increasing oxygen and energy available to our body, also enhancing alertness.

When it is put like that, the changes begin to sound quite positive. This is a far cry from the negative way in which students usually experience such physical changes. That said, we need to be aware of the full range of physical side effects of heightened sympathetic nervous system activity, such as a dry mouth, pronounced heart beats, and dizziness, all of which are typically perceived as unpleasant. Armed with the knowledge and understanding of such changes in the body, however, can rapidly change a student's mindset. This has been Tom's experience when speaking to teenage Maths students, observing light bulb moments when teens begin to show awareness of the physical changes they experience in relation to their own Maths Anxiety.

Breathing Techniques

It has long been known that mindfulness, the act of being fully present in your surroundings and aware of where you are and what you are doing, and relaxation-based strategies are effective in tackling anxiety in a general sense. There have also been attempts by researchers to systematically test these strategies in the context of Maths Anxiety, which has resulted in several different techniques being explored.

A well-researched intervention, laid with the foundations of mindfulness and relaxation, are breathing techniques. These strategies ask an individual, or a group, to practise different breathing methods to alleviate any feelings of stress and anxiety, which have been found to be effective in reducing levels of Maths Anxiety. There are different strategies available though, with varying effectiveness.

> Samuel and Warner (2021) researched the impact of conducting controlled breathing exercises at the start of every lesson over an academic term for 40 university students in the US. They reported that this regular use of focused breathing allowed the students to be aware of their surroundings, and that their Maths Anxiety towards solving Maths problems was reduced.

Focused breathing involves an individual slowing down their breathing, becoming more mindful of their breathing, and decreasing the amount of oxygen that the body takes in, leading to a sense of relaxation. This prompts the individual to be seated in a relaxed position and avert their attention to the mechanism of breathing, such as inhaling and exhaling, and focus their attentional resources on this. If an individual feels that their mind is wandering, they are reminded to avert their attention back to the sensation of breathing.

> Brunyé et al. (2013) tested the effect of different breathing techniques to reduce Maths Anxiety in 36 university students. Each breathing exercise lasted for 15 minutes and was modelled on Arch and Craske's (2006) work. Participants were asked to listen to a recording, which told them to sit in an upright position, rest their hands on their lap, relax their shoulders, have their head upright, and rest their feet flat on the floor, either looking downwards or having their eyes closed. The focused breathing exercise worked to increase the participants' attentional focus on the sensation of breathing, such as inhaling and exhaling. The unfocused breathing exercise asked participants to *"simply think about whatever comes to mind."* Results showed that the Maths performance of high maths anxious students was closer to the Maths performance of those who were low in Maths Anxiety when the focused breathing strategy was utilised, compared to the unfocused breathing strategy.

Various other breathing techniques aim to reduce Maths Anxiety by providing patterns of breaths for the individual to follow, such as the 5/7 breathing technique. This is where individuals, again seated in a relaxed position, are asked to hold their breath for 5 seconds, and then breathe for 7 seconds, in a continuous cycle. Following a cyclical pattern of breathing in this manner averts the individual's attention to their breathing, increases mindfulness, and reduces thoughts and feelings related to anxiety and stress.

> Johnston-Wilder and Marshall (2017) used the 5/7 breathing technique as one approach to reducing anxiety in maths anxious students in a support centre. They found that, as part of a bigger intervention, this breathing technique had positive effects and calmed anxious individuals when they self-identified that they were in the "anxiety zone." It was suggested that this was due to the extended out-breaths having a direct impact on the amygdala, associated with the stress response.

Breathing techniques can be implemented with ease, as teachers, and parents at home, can access online content to learn about the theory behind this and how to successfully use these relaxation strategies to reduce Maths Anxiety. Providing the anxious individual with these tools is key; modelling the strategies and explaining why they're useful, and how they work, can provide an individual with a useful strategy that they can implement independently.

These strategies can be discussed, learnt, and utilised in one-on-one situations but can also be within a group setting, or a whole class environment. How and when to utilise these strategies is very dependent on the teacher and how it would work within their classroom environment. We have seen some teachers embed these within their classroom culture successfully, using breathing techniques as part of mindful discussion at the beginning of the day, or at the start of Maths lessons, or within weekly lessons focusing on well-being. There are endless possibilities as to when this strategy can be taught, and used, by individuals, but it is important that the maths anxious person is engaged with using the strategies for purposeful effect. As a low-energy, low-time strategy that has been shown to improve Maths Anxiety, it is definitely worth reflecting on how this could be implemented within your environment to support those experiencing Maths Anxiety.

Ezra's Experience

Ezra was prone to panic during Maths tests. A key feature of this was rapid, shallow breathing, and this exacerbated his panic even further. Ezra's teacher decided to spend some time talking to the class about breathing effectively. This seemed strange to the class at first as they had never really thought about their own breathing. However, things started to make sense when the teacher spoke about using the diaphragm effectively during inhalation, while keeping the chest and shoulders relaxed. The teacher gave carefully timed reminders so that the class could check in on their breathing, practising their diaphragmatic breathing. This was usually at the start of a Maths lesson, or when more challenging Maths problems or new topics were presented. The night before his next Maths test, Ezra practised the new breathing technique and made sure that he did this again throughout the morning of his test, at the start of his test, and at regular intervals throughout the test. He was pleasantly surprised at just how effective this turned out to be at helping him reduce the sensation of panic.

Relaxation Strategies

Further strategies related to increasing mindfulness and relaxation to reduce Maths Anxiety have been less researched compared to breathing techniques but are interesting to explore. Activities such as colouring, listening to music, going for a walk, or using virtual reality, as a few examples, have been found to reduce an individual's levels of Maths Anxiety, due to increasing an individual's mindfulness and relaxation.

> Salazar (2019) reported a positive effect of colouring mandalas on Maths Anxiety. They instructed a group of undergraduate students to colour in mandalas, while the control group was instructed to simply doodle. The researcher found that Maths Anxiety was significantly reduced in the colouring condition, while no significant reduction was seen in those who simply doodled. This emphasises the benefits of this particular type of colouring as a therapeutic technique, inducing both cognitive and physical calm.

Most of the research into relaxation and mindfulness strategies has been tested at a group or class level, which shows how whole-class approaches can be effective. It is worth bearing in mind, however, that there will always be individual preferences when it comes to engagement in specific non-Maths activities, whether that is colouring or listening to music. Such preferences need to be taken into account so as not to produce counterproductive effects as a result of other factors, such as boredom or a lack of enjoyability. For example, encouraging a maths anxious student who hates colouring to partake in mindful colouring activities is unlikely to have a positive effect. Therefore, it is again important to remember that knowing the students within the classroom will allow you to design and select purposeful and relatable strategies that are more likely to successfully address Maths Anxiety.

> Davis and Kahn (2018) used virtual reality in an attempt to calm maths anxious individuals. The participants were presented with calming scenarios in virtual reality, which resulted in reduced Maths Anxiety and improved Maths performance, compared to a control group.

One argument sometimes put forward about why relaxation techniques have a positive effect on Maths Anxiety is that certain activities simply act as a distraction from anxious feelings. In the case of music, however, stimulative music has not been found to produce the same beneficial effects as sedative music. This suggests that the distraction argument does not hold.

> Gan, Lim and Haw (2016) showed how attending to sedative music during a Maths task can be beneficial for students. They took a range of physiological measures and also measured state anxiety and Maths Anxiety before, during, and after a Maths task. Students were assigned to one of three groups: sedative music, stimulative music, or no music. The results showed that listening to sedative music led to a significant reduction in state anxiety and blood pressure. The researchers explained the findings using a perception-to-physiology model, in which the sedative music first reduced perceived anxiety and this, in turn, supported physical relaxation over time. In other words, students might acknowledge the music as psychologically relaxing and then, over the course of say 20 minutes, they also begin to feel physically relaxed.

Using these relaxation strategies within the classroom is relatively straightforward, as long as they are related to the individuals' preferences and delivered with purpose. It is important to explain to the individual that the reason behind them listening to relaxing music, or going for a walk, for example, is so that they can become in tune with their emotions and heighten their understanding of ways in which to overcome their feelings of anxiety. Time again is another aspect to consider, weighed against the impact of the strategy. For example, if an individual finds going for a walk helps to reduce their anxiety, how can this be successfully managed so that they do not miss the majority of their Maths lesson? How can this be monitored so that the relaxation strategies don't turn into another form of avoidance? These are questions which need to be carefully considered when implementing these strategies within the classroom, and at home, to support individuals within mathematical situations which evoke their feelings of anxiety.

Sharing these strategies with parents would be useful in increasing their effectiveness. Not only by talking with them about their child's interests and what they do to relax at home, which will provide the teacher a wealth of information about the child, but through aligning strategies in both the home and school environments as well. As Maths does not just exist within the classroom, sharing these relaxation strategies may provide support for maths anxious individuals in a wide range of mathematical situations that evokes their anxiety.

An important consideration when exploring relaxation strategies is resourcing, as it is unlikely that there are multiple virtual reality headsets available for maths anxious individuals to use when they are feeling anxious, for example. So, while these are, in general, quick and efficient strategies that can reduce levels of Maths Anxiety, reflection is needed as to how to embed these within the classroom, or at home, and ensure that the maths anxious individual is aware of how to use these strategies successfully.

> Johnston-Wilder and Marshall (2017) found that listening to calming music, as one of many relaxation strategies, can lessen an individual's anxiety response to a mathematical situation. They did offer individuals a range of strategies, so the success of this may be linked to the individual's interest and personal preferences, as they chose calming music as a method of relaxation over other suggestions.

What underlies these various approaches though is an attempt to regulate emotions. Mindfulness and relaxation may offer a relatively quick and easy way to avoid panic and reduce high levels of arousal. However, in the context of Maths Anxiety, particularly based on what we know about the cognitive aspects of it, it is likely that other strategies need to be considered alongside them.

Systematic Desensitisation

The idea behind systematic desensitisation to address Maths Anxiety is that people are gradually exposed to the thing that makes them anxious; it usually involves relaxation as part of the process.

Take spiders, for example. Someone who is afraid of spiders is likely to feel a greater level of anxiety in response to being faced with a real-life spider compared to, say, seeing a cartoon spider on the television. That's not to say they wouldn't feel some degree of anxiety at being exposed to a spider in whatever form it is, but addressing a low-level anxious response could be more manageable. Based on this, learning how to manage anxiety in response to seeing a cartoon spider could be a good place to start. Thus, this would involve some form of anxiety management technique, which is often relaxation based, while being exposed to the cartoon spider. Only once the person feels able to regulate their emotional response at this lower level of exposure would they move on to the next level of exposure. Continuing with our example, this could be a photo of a real spider, followed by a video of a real spider, followed by being in the same room as a real spider while being separated from it in some way, and so on.

> Akeb-Urai, Kadir and Nasir (2020) examined the effectiveness of systematic desensitisation on Maths Anxiety and Maths performance in 65 college students in Malaysia. They reported a significant reduction in the level of self-reported Maths Anxiety, and an increase in their Maths performance.

Through utilising this approach, gradual exposure to Maths could follow a similar suit; if someone is anxious about learning and using Maths, gradual exposure to this may assist in reducing the anxiety they experience. This could involve a gradual increase in difficulty level, integrating relaxation along the way. However, for some students, it might be beneficial to learn how to manage their anxiety in situations associated with Maths learning, rather than being exposed to any actual Maths. Therefore, it could be a case of starting with exposure to Maths-based contexts: simply viewing Maths problems or listening to someone talk about Maths. Only when they feel comfortable would they move on to actual mathematical problem-solving.

> Sheffield and Hunt (2007) used systematic desensitisation as an approach to improve Maths Anxiety in primary school children. The one-hour session, modified from Meichenbaum's (1977) cognitive modification programme, involved practising diaphragmatic breathing, the use of imagery for anxiety reduction, and in situ desensitisation. The latter consisted of graded exposure to increasingly difficult Maths problems, while practising diaphragmatic breathing. Pupils were also instructed to practise this at home. Despite some baseline differences between the intervention group and the control group, the intervention resulted in a reduction in Maths Anxiety and an improvement in Maths performance.

For teachers to implement this in the classroom, or at home as part of a home learning programme, would be quite challenging to do correctly. Research has shown that to be effective, multiple sessions of systematic desensitisation are needed, with gradual exposure occurring at the correct pace tailored to the individual. This is time-consuming and can't occur during a whole class lesson, so consideration will be needed for scheduling this in for the individual, as well as finding a quiet, relaxing place for this to occur in.

The difficulty further lies in ensuring that the approach is utilised correctly, with specialist training being required to ensure the safeguarding of the students and that they are not placed under any undue stress or pressure through this process. This can be quite costly and again requires additional time for training measures. While it has been shown to be an effective strategy to improve Maths Anxiety, systematic desensitisation is not a freely available intervention to be used by teachers within schools, making it a niche branch of intervention to support Maths Anxiety, perhaps best utilised by trained educational psychologists.

Expressive Writing

It is surprising to hear of the lack of open discussion about anxiety, and emotions more generally, about Maths within the classroom. While it perhaps shouldn't be so surprising given the limited resources available, such as staff and time, it makes us question how long negative feelings towards Maths can bubble along under the surface, and what impact this can have. On top of this, students might not always feel inclined to express how anxious they feel, which might be for a variety of reasons. For example, they might not recognise their feelings as being anxiety, or they might feel afraid of the consequences of expressing how they feel, or they might struggle to articulate their feelings.

> Sammallahti et al. (2023) conducted a meta-analysis and highlighted the benefit of both cognitive-based and emotion-based interventions for reducing Maths Anxiety and increasing Maths performance. They found that interventions based on motivation alone were not helpful when it comes to Maths Anxiety and performance.

Before they become debilitating, it is important to teach students how to understand their emotions, express them, and regulate them. A common activity proposed by therapists to achieve this is for their anxious clients to keep a journal. This often involves writing down one's thoughts and feelings, usually at the end of the day when there is a bit more time for relaxation and reflection. This can help with identifying patterns, for example, which thoughts and feelings occur the most frequently, but also at what times, such as a response to particular actions or experiences. Journals might then offer anxious people a way of understanding their own thoughts and feelings. More simplistically though, writing offers a way of externalising thoughts, like a way of decluttering one's mind.

While researchers haven't adopted a full-on journal approach to tackling Maths Anxiety, they have tested the concept of expressive writing. This strategy, also known as writing therapy, asks individuals to write honestly and freely in order to make sense of their thoughts and feelings. This is often about a particularly distressing event, or situations that cause anxiety or stress. For example, a maths anxious individual may write about their feelings of anxiety related to an upcoming Maths test, or their anxiety related to using algebra correctly. Expressive writing is thought to be an effective strategy to combat Maths Anxiety, as it helps individuals regulate their emotions by aiding in cognitive restructuring and directing their attention to their emotions.

> Park, Ramirez and Beilock (2014) tested the idea that a brief expressive writing task would reduce intrusive thoughts and improve working memory. They asked one group of university students to briefly write about their thoughts and feelings about a Maths test they were about to complete. Another group of students simply sat quietly without writing. In this latter group, those who were high in Maths Anxiety performed significantly worse on the Maths test compared to students who were low in Maths Anxiety. Supporting the researchers' hypothesis, this difference was reduced in the expressive writing group.
>
> Taking it a step further, the researchers also looked at the type of words expressed by the students. In the highly maths anxious students, they found that the greater the use of anxiety-related words, the better the Maths performance was. This adds credence to the notion that expressive writing frees up working memory in those who are highly maths anxious. To add to this, the relationship between frequency of anxiety words and performance was only observed for Maths problems that placed greater demand on working memory.

As there have been many positive effects of expressive writing found on Maths Anxiety, it sounds like it should be the go-to strategy to support maths anxious individuals. Arguably, any form of expressive writing has the potential to support emotion regulation to some extent, regardless of what the writing is about, due to providing a distraction from intrusive thoughts and worries specific to Maths. However, not all studies have reported such positive effects.

This may be related to the age of individuals utilising expressive writing. Writing is a skill in itself and one might expect adults, especially highly educated ones, to have the skill to articulate their thoughts and feelings in a way that young primary school children simply can't. Therefore, a fundamental question to ask is whether certain interventions work for students of different ages. After all, it is hard to draw too many similarities between students in early primary school and adults. It could be that expressive writing is more suited to students in secondary school and beyond, rather than primary school.

Another factor that might influence how effective expressive writing is, is time. In some studies, the intervention focusing on expressive writing took place over a longer period of time, such as a week. On the other hand, some studies focused on one off instances of using expressive writing, such as before a Maths test. In this latter case, there would have been no opportunities for other things to have affected anxiety levels in the run-up to the Maths test, and it is the one-off instances that have shown to have a positive effect upon Maths Anxiety. In other words, maybe the benefits of expressive writing are time-limited and depend on the activity taking place just prior to Maths testing, rather than prolonging the process.

> Walter (2018) found no differential effects of expressive writing on Maths Anxiety or performance according to groups of primary school children who were low or high in Maths Anxiety. In this instance, the intervention involved writing about Maths worries prior to

> a range of Maths tasks over the course of a week. The activity in the control condition also differed to the one used in other studies, such as Park, Ramirez, and Beilock's (2014). This time, rather than sitting quietly and doing nothing, the students in the control group were still instructed to write expressively, but on a topic of their choice.

We can also look at the expressive writing task itself. Is the content of an individual's writing important? It is possible that writing exercises, regardless of their content, can have a general calming or cathartic effect, which benefits maths anxious individuals. If the expressive writing task instructions explicitly refer to previous events or experiences, to what extent does that specific mention have on a student's thoughts and feelings? After all, we know that the way in which a student appraises their previous experiences of Maths is associated with how anxious they are about the subject.

> Hines, Brown and Myran (2016) instructed one group of 14-19 years olds to write about their feelings concerning Maths, the state tests (the study was conducted in the US), and about school. The control group wrote about their plans after high school, perceptions of teachers, and their favourite time of year. Students wrote for 15-30 minutes a day, for three days. The researchers found that Maths Anxiety was reduced in both groups, irrespective of the exact topics the students wrote about.

There are so many potential ways in which writing activities can be designed and implemented, so we urge teachers to think carefully about this. Expressive writing tasks do come with a warning in that, by their nature, the content could be quite sensitive. An outpouring of emotions on paper, or screen if done digitally, might be overwhelming for some students, so the context of the activity should be thought through, along with what questions will be asked. Consideration also needs to be given to what plans are in place if a student becomes upset, and whether students are more or less likely to express their emotions if they are in the presence of others, or if there is potential for peers or teachers to view what they have written.

Charlie's Experience

Charlie would often feel overwhelmed with worry as she sat down to revise for her upcoming GCSE Maths exam. Her teacher suggested that, before she even starts a revision session, she writes down a few words on her phone that describe her worries, then deletes what she has written after she has spent some time revising. After a week or two, Charlie's teacher asked her about this experience, to see whether Charlie found it helpful (perhaps calming), whether it made things worse (encouraged even greater worry), or made little difference. The teacher was unsure which way the conversation would go and was prepared to suggest stopping the activity if it was not helpful. It turned out that Charlie found the exercise to be very useful, both in terms of reducing the number of intrusive thoughts as she was revising, but also recognising that she repeatedly wrote down the same worries.

It should also be noted that expressive writing interventions are usually based around the concept of cognition, having a focus on what individuals are thinking, with the assumption that there is then a direct link between thoughts (and processing those thoughts in some way, through writing them out) and the student's test performance or self-reported anxiety. Writing-based interventions for Maths Anxiety often neglect to consider the possibility of underlying ways in which the processing of thoughts and feelings might indirectly impact certain outcomes, such as performance or anxiety. Instead, the focus is usually on the simple act of writing down one's worries, thoughts, and feelings.

> Ganley et al. (2021) tested the effectiveness of an expressive writing intervention, amongst others, with University students in the US. They gave the following instructions to students:
>
>> Please take the next 5 minutes to write as openly as possible about your thoughts and feelings regarding the Maths problems you are about to perform. In your writing, I want you to really let yourself go and explore your emotions and thoughts as you are getting ready to start the Maths test. You might relate your current thoughts to the way you have felt during other similar situations at school or in other situations in your life. Please try to be as open as possible as you write about your thoughts at this time. Remember, there will be no identifying information on your essay. None of the experimenters can link your writing to you. Please start writing.
>
> They found that expressive writing carried out just before a Maths test actually increased state anxiety; the opposite to the desired outcome. The researchers suggested that the task actually activated anxiety in students, as it got them to focus more on the upcoming Maths test, which generated more anxious thoughts or feelings.
>
> This study focused on state anxiety (how anxious a person feels at the current time, in a particular moment) though, rather than general Maths Anxiety. The researchers also asked students to report on their level of state anxiety retrospectively. Usually, researchers will ask participants what their state anxiety is right at the time of interest – in this case, it should have been before the Maths test took place. Instead, the researchers asked students after the test to report on how they felt before the test. This can be problematic for several reasons, ultimately questioning whether the students' reporting was reliable and valid.

One suggestion is that a strategy such as expressive writing might support emotion regulation via physiological effects, producing a soothing or calming response, for some people at least. If that is the case, then those people might see benefits of such an intervention, whereas for others, it might have the reverse effect, being activating or arousing. Therefore, the potential range of individual differences in how something like expressive writing might operate could be muddying the waters when we look at a simple, direct link between an intervention and outcomes, especially across whole groups.

> Ramirez and Beilock (2011) tested the usefulness of expressive writing in secondary school students just prior to a high-stakes Maths test. Students were told that their performance would impact whether others received a reward, and they were told that they were being observed, to create feelings of pressure. They found that expressive writing about the upcoming test had a beneficial effect on the students' Maths performance.

A final aspect to consider is that Maths Anxiety often affects a student's Maths performance when there is an increased feeling of pressure. This therefore adds yet another layer to things – that is, whether the nature of the test, including the degree of pressure and perceived consequences of underperforming, affect whether expressive writing has a negative, neutral, or positive impact on students.

Using expressive writing in the classroom could be used at various times, with either an individual, a small group, or the whole class. Discussing what expressive writing is, and its benefits, could be a useful tool to provide to all students, as well as those who are maths anxious. This strategy, once learnt by an individual, can be used independently. Expressive writing has been shown by research to be most effective just before a stressful situation, such as a Maths test, so this may be a good place to start.

Teachers could evaluate the effectiveness of the strategy on the maths anxious students within their classroom and identify the value of using it ahead of anxiety-evoking situations, or on a more regular basis. For example, at the beginning of each day, students could be provided time to use expressive writing as a way to offload their intrusive thoughts so that they can fully focus on the day ahead. However, again, it is important to reflect on the individual nature of students. If some students really dislike writing, expressive writing might not be an effective strategy to reduce their feelings of Maths Anxiety. Instead, it may be seen as another pressure, and simply another activity that needs to be completed. It is also important to assure students that their writing is personal and confidential to them, as this may affect the outcome and effectiveness of using expressive writing as a strategy to reduce Maths Anxiety.

Psychodrama Group Therapy

Taking a somewhat similar approach to expressive writing, at least from the point of view of facilitating the expression of thoughts and feelings, psychodrama group therapy is a strategy to support those experiencing Maths Anxiety. The idea with this approach is that it provides a safe place for students to re-enact previous Maths experiences that they regard as negative. For example, they may re-enact being anxious when adapting a recipe for a family meal, or feeling embarrassed when they got a Maths question wrong in front of their class.

Through psychodrama group therapy, students build up to the end result of creating and acting out plays, which support their understanding of their emotions and the situations which they find anxiety-inducing. If students are asked immediately to create a dramatic piece reflective of their Maths Anxiety, this could be quite triggering and damaging to their well-being. This is why the process needs multiple sessions over an extended period of time, such as an academic term. Within this though, there needs to be an agreed level of mutual respect regarding the differing extents to which students express their feelings. It is also essential that there is an agreed level of privacy between all who are involved to create a groundwork of trust and openness.

> Dorothea (2016) reported on the use of psychodrama group therapy for the treatment of secondary school students who were high in Maths Anxiety. Engagement in the psychodrama group therapy led to students feeling more able to approach mathematical situations in a relaxed way. They also reported having less negative or worrisome thoughts about tests as a result of the intervention. As the sample size was quite small (less than ten students) and the data analysis was limited to descriptive statistics, it is difficult to draw too many firm conclusions. However, the findings do suggest that there could be some value in psychodrama group therapy as an approach to tackling Maths Anxiety.

Using psychodrama group theory as an intervention for Maths Anxiety can be quite resource intensive, as well as being time-consuming. This is problematic for a lot of schools looking for quick, and cheap, strategies, but it might be something that is considered for students with particularly high Maths Anxiety. The intervention is also likely to require someone with suitable expertise to facilitate the sessions, and as with other expressive techniques, such as expressive writing, there are obvious risks associated with the activities activating anxiety, as well as other risks inherent in sharing emotion as part of a group. Therefore, if you are to use this strategy within your classroom or school, ensure that whoever is delivering it is experienced in using drama as a teaching tool, and that support is in place for students' well-being should there be any issues that arise from using this strategy.

As with general methods for creating a relaxed and calm Maths learning environment, strategies such as psychodrama group therapy might not do enough to directly address the underlying Maths Anxiety. Therefore, attempts to support students in understanding their own Maths Anxiety, and open up the conversation, might be needed.

Stories and Storybooks

Starting a conversation about Maths Anxiety, or thoughts and feelings about Maths generally, isn't always easy. Individuals might have difficulties in knowing where to begin, or how to articulate their feelings, and it might be the case that they do not particularly want others to know how they are feeling. A relatively new, and still developing, approach to addressing Maths Anxiety is the use of stories or storybooks. Rather than the student writing about themselves, they can read about the experience of others.

> Passolunghi, De Vita and Pellizzoni (2020) devised a combined approach of a range of individual strategies to reduce Maths Anxiety in primary school children. One component involved the use of stories. Children read a short story called "Maths has trapped me" and were instructed to discuss the character's feelings. Not many details are provided by the researchers and, with studies that combine multiple strategies, it is difficult to understand whether changes in anxiety are down to specific strategies or their combination. Nevertheless, a storybook approach, especially with younger students, sounds promising for initiating conversations about Maths Anxiety (whether that term is actually used or not), potentially normalising Maths talk.

While using stories and storybooks can be used as a targeted strategy to support students who are particularly maths anxious, it also has the benefit of encouraging wider discussion and the sharing of ideas and feelings, creating a more inclusive Maths classroom environment. These stories could be read as a whole class, in groups, or one-on-one, and shared with parents so that they can be read at home. Many follow-up activities could also be used in order to embed the student's comprehension of the stories and better understand the experience of Maths Anxiety and how to deal with this. For example, interviewing characters from the stories to identify how they are feeling, writing a diary entry through the eyes of the character, or drawing a picture that represents the character's emotions through the story.

> Petronzi, Schalkwyk, and Petronzi (2024) produced a story book that belongs to a wider series of research-informed children's books: *"Whoopsie Doodle Little Noodle."* Their Maths Anxiety storybook focuses on the protagonist (Noodle) who experiences worry as she encounters three Maths-based situations at home. With the support of her family, she learns how to approach Maths problems and gradually grows in confidence. The book was used as a Maths Anxiety intervention in the classroom in an attempt to normalise Maths talk and to support emotion regulation of young school children, aged between six and seven years. Rather than simply providing children with the book to read, the procedure involved reading with the student on a one-to-one basis, asking them questions as they went along. This was designed to encourage reflection on the story itself but also to encourage discussion of the individual's own thoughts and feelings about Maths. Through this process, the students were able to link the experiences of the characters to their own emotions. The researchers suggest the storybook approach can complement a school's Maths scheme of work, rather than taking up time or resources.

Animal Therapy

The support of animals through therapy has seen a surge in schools in recent years, with their calming presence having positive effects on students' well-being and feelings of stress, worry, and anxiety. Building on a concept known as the "human-animal bond," the desire to interact with animals and form bonds can produce a calming effect on an individual. These animals can vary, all the way from horses to cats and rabbits, but therapy dogs have been heralded as particularly effective in reducing anxiety, increasing overall well-being and emotional and cognitive function in students.

> Kirkpatrick (2020) conducted a two-week pilot study to investigate whether daily sessions with a therapy dog had a positive impact on student anxiety and attitudes. After the two weeks, students overall had improved self-belief and fewer anxieties. Although this was in relation to anxiety in social situations, rather than Maths, this highlights the positive impact that therapy dogs can have on anxiety.

Focusing specifically on the impact of animal therapy on Maths Anxiety, researchers have shown that therapy dogs can have a positive effect and reduce levels of Maths Anxiety. There does need to be more research into this to substantiate it as a reliable strategy to use to combat Maths Anxiety, but the findings are promising.

> Soll and Petronzi (2023) carried out a therapy dog intervention for Maths Anxiety in primary school children. This included two 15-minute sessions with a therapy dog, named Ridley, in the Maths class, and children completed Maths problems in a dog-themed workbook. This was designed to be fun, but the Maths questions and problems were also framed around the therapy dog. Positive messages about Maths were also framed around the therapy dog, such as "Mistakes are Riddley's superpower!". Children were also encouraged to engage in self-reflection and cuddle the dog whenever they experienced worry about Maths. This approach not only helped them identify their worries but also supported emotion regulation, learning how to manage their emotions and worries effectively. Findings showed that students' self-reported Maths Anxiety was lower after engaging in the intervention.

Animal therapy as an approach does offer another option in tackling Maths Anxiety, but it does have obvious drawbacks in terms of the availability of therapy dogs and experts, and the possibility that some children may have an allergy or be fearful of dogs. On the whole though, where it is pragmatic to do so, therapy dogs seem to offer a valuable solution, even if a more targeted approach is taken, focusing on students who are particularly high in Maths Anxiety. Of course, rabbits and cats have been shown to be great therapy animals as well but, only therapy dogs have been explored in relation to their effect on Maths Anxiety.

> Vokatis (2023) shared how dog therapy had a positive effect on children who found Maths difficult, as dog Carmel helped with their Maths lessons and was even called the "math checker." Findings were "overwhelmingly empowering" and children enjoyed participating in the Maths sessions.

If using therapy animals may not be applicable in your school setting, you may instead consider some related approaches, yet there are questions over their suitability and effectiveness. For instance, one therapy that has been used in educational settings is the use of puppets in lieu of animals. This approach might mitigate some of the challenges that exist with dog therapy or related interventions involving animals, but it has yet to be tested with the specific aim of reducing Maths Anxiety.

> **Rudy's Experience**
>
> Rudy felt anxious during his Maths lessons, much more so than his classmates. His teachers were unsure why his anxiety in this specific subject was so high when he did not display such anxiety in his other subjects at school, and his attainment showed that he had a good grasp of Maths. Fortunately, a local charity offered the services of a therapy dog named Poppy. The dog, its trainer, and a teaching assistant worked with a small group of pupils, including Rudy, twice each week. Poppy became the focal point for Maths problems and the students would offer suggestions, strategies, and proposed solutions as though Poppy was actually offering these. This seemed to minimise the students' fears of being judged, encouraging a much greater level of discussion compared to what normally took place. Rudy found this very comforting and found his anxiety in Maths classes, even without Poppy being present, slowly improved.

Cognitive Restructuring

Strategies based on cognitive restructuring have been shown to be successful in reducing Maths Anxiety. Cognitive restructuring is based on supporting individuals in noticing their negative thinking patterns, and what effect they are having on them, before helping them to change this and redirect them towards more positive outcomes. This approach in relation to Maths Anxiety addresses faulty beliefs concerning Maths, which could include unhelpful, erroneous, or irrational thoughts about an individual's own Maths ability or performance, or even about Maths more generally as a subject. These negative thoughts could be everyday intrusive thoughts, such as *"I'm not good enough,"* to cognitive distortions, which lead to a distorted view of reality, including catastrophising: "Everything is riding on this Maths test, everyone is going to know I failed and I will never get a job."

A key part of cognitive restructuring is self-monitoring, whereby the anxious individual needs to be able to identify patterns behind the negative thoughts and beliefs, in order to understand them better and overcome them. For example, is someone maths anxious mostly when it comes to learning specific mathematical content, or in group situations, or in relation to assessments? The patterns that emerge through discussing these negative thoughts and beliefs can be very useful in understanding the person's experience of Maths Anxiety better, and helping them be free from these hindering beliefs.

> Hunt and Maloney (2022) found that a higher level of Maths Anxiety is related to a greater likelihood of appraising one's previous Maths experiences negatively.

A way in which to support cognitive restructuring is to record the intrusive, negative thoughts. This might be useful in the form of a journal, or an electronic log, made by the anxious individual. While they do not need to share this with you, it would be useful to spark discussion of any patterns, similarities, and ongoing issues that they begin to notice as the full picture emerges within their note taking and reflections. For example, a maths anxious individual may identify that they are regularly catastrophising about being asked a question in front of their peers in a Maths lesson, and getting it incorrect. This may spark a discussion about why this is a key, negative thought that they have, and what evidence they have to support that the negative event they have anticipated will occur. If it did happen, what would be the worst thing that could happen as a result?

> Asikhia (2014) investigated the effect of cognitive restructuring upon Maths Anxiety levels of 180 secondary school students in Nigeria. The experimental group received cognitive restructuring training, while the control group received a placebo treatment. They found that the group who received cognitive restructuring training had a greater reduction in Maths Anxiety compared to the control group.

Supporting the individual in questioning these thoughts, once they have been able to identify them, may help them reason whether they are based on fact or emotion, and whether they are accurate and evidence-based. For example, if a maths anxious student is of the belief that they will never be able to use fractions and have identified that their negative beliefs are related to using and learning about fractions, what evidence do they have of this? Could they test this belief? What do their previous experiences tell them? This is particularly useful if the individual's automatic response to mathematical situations is to catastrophise, as it allows them to identify alternative possibilities, which are not always the worst possible outcome, and reason about the likelihood of different events occurring.

Finally, it is important to discuss with the individual the impact of these thoughts upon them. Are they benefitting from these negative thoughts? Does this impact how they engage in Maths lessons? How is this limiting them? How does it positively impact their goals? Again, opening this level of discussion is thought-provoking for the individual, but key to cognitively restructuring their negative beliefs about Maths. This might be an early stage in the person adopting alternative, positive thoughts that replace the original negative thoughts, including positive affirmations such as *"I can do this."*

> Parameswara, Utami, and Eva (2022) aimed to test the effectiveness of cognitive restructuring as a strategy to reduce Maths Anxiety in six secondary school students. They conducted three sessions of cognitive restructuring with the participants, each lasting between 1 and 1.5 hours long. While the sample size was small, they found that cognitive restructuring was effective in reducing Maths Anxiety. They recommended that these sessions should be carried out by professionals, such as psychologists, and that more sessions may have led to increased effectiveness.

Cognitive restructuring has been shown to be beneficial in changing negative affect towards Maths and an associated lack of self-belief, reducing levels of Maths Anxiety. This strategy requires the time and resourcing to be available for regular one-on-one discussions, to truly allow the individual to interpret and understand the impact, and source, of their negative thought patterns. While cognitive restructuring can be done in small groups, the individuals' thoughts and experiences may need to have a similar basis in order to be effective and meaningful.

A teacher may be able to pick up on some negative comments within the classroom and support the student directly there and then with some knowledge of cognitive restructuring, but it is unlikely to be consistent and meaningful enough to lead to positive change. Similarly, parents may be able to identify any negative comments used at home related to Maths, but without the training and understanding required, the impact that they can have upon this may not be effective. In order to deliver purposeful cognitive restructuring interventions, similar to systematic desensitisation, specialist training would be required in order to deliver it successfully, or it would need to be delivered by a professional such as a psychologist with the necessary training. Again, this protects the students and ensures that their well-being is of the utmost priority.

Once the tools of cognitive restructuring have been taught, the individual would be able to subsequently complete this on their own, leading to fewer sessions with a professional being required, with potential follow up sessions when the individual feels that they require one. Importantly, this does rely on the individual being old enough to self-identify the patterns of negativity, and to be able to record and discuss these. Therefore, it may be more successful as a strategy with secondary school students, or older.

Reappraisal

Interestingly, researchers have used the principle of reappraisal to design and test interventions which address Maths Anxiety. Reappraisal refers to facilitating the way in which a person can change the way a situation, event, or stimulus, is interpreted by them, to result in a more positive interpretation. This might even mean flipping certain thoughts on their head and reinterpreting a certain feeling as being beneficial, which might be the opposite to how the individual person would normally perceive it.

> Jamieson et al. (2016) established two groups of adult students. Those in the reappraisal group were informed that increased arousal experienced during testing is not harmful, and that arousal improves performance as stress responses help us cope with demanding situations. Students in the other group were instructed to ignore any stress they experienced, including negative thoughts. The first group demonstrated greater improvement in performance from one Maths test to another, compared to the second group, and their self-reported Maths Evaluation Anxiety reduced too. These effects were long-lasting, with the intervention leading to higher grades in the reappraisal group compared to the control group by the end of the semester.

Studies that utilise reappraisal to reduce feelings of Maths Anxiety demonstrate the power of how students appraise their own feelings, potentially reappraising how they feel with long-lasting, positive effects.

> Pizzie et al. (2020) conducted a similar study to Jamieson et al. (2016) with another group of adult students. They gave some the option of reappraising, and others formed a control group. Reappraisal was shown to attenuate the negative effects of Maths Anxiety on Maths performance, once again supporting the value of facilitating a change in an individual's thoughts about Maths and their responses to it.

Research hints at just how unhelpful many individuals' thinking can be when it comes to appraising their responses to Maths. Now, it might be argued that such thinking is not context-specific. However, we often experience even a mild stress response to demanding situations without interpreting the response as a threat. In many instances, we might actually perceive the response as enjoyable, in the case of being flung around on a roller coaster, for example. Yet students, and the general population at that, often fail to acknowledge the incongruence in how they appraise their responses across different situations.

In the many discussions with individuals on this topic, Tom has found that people are very good at identifying previous challenging situations they have experienced in which they can talk about heightened arousal in a positive way. The difference tends to be that the situation had not automatically been regarded as a threat at the time. It is often helpful to initiate a discussion with individuals about sports, hobbies, clubs, family games, or any other similar situation that might, from time to time, have resulted in real or perceived competition or evaluation for the individual. These topics have the advantage that they are relatable to most people. Through exploring these experiences further, individuals come to realise that there is often an advantage to heightened arousal, and that a challenge without any overt danger can lead to an adaptive response. From that perspective, it is perhaps easier to see how practice at subsequently applying this thinking to mathematical situations can result in a step-change in how a person appraises their feelings.

> Ganley et al. (2021) conducted a study in which students simply read the words "try to get excited" on a computer screen before beginning a Mathematic test. This approach resulted in no change in anxiety or Maths performance.

There is an element of control and safety in taking this approach. Similarly, one activity that can modify an individual's appraisal of Maths situations for the better is to encourage them to give advice to others. Firstly, they could imagine a non-mathematical context and consider what advice others might give. A typical example might be where the individual puts themselves in the place of a sports coach and imagines a scenario where a sportsperson is about to perform, for example, at a cup final in a football match. The sportsperson, in this case a footballer, might appear anxious; they might verbally describe their feelings of nervousness and worry, and they might show observable signs of anxiety. The individual is then encouraged to think about the advice they would give to the footballer. More often than not, they give really good advice. They offer the footballer support, usually adopting very positive language too, encouraging the footballer to use their heightened arousal to their advantage, while also affirming positive beliefs, such as:

> "You've got this!"
> "I believe in you!"
> "Remember what you did in training!"

The activity might then move to a Maths-based scenario, whereby the advice turns to supporting a maths anxious friend or relative ahead of, say, a Maths test. Students will usually provide the same advice they gave in the previous scenario, in this case, to the footballer. If they are asked to put themselves in place of the maths anxious student, and to then consider what advice they would give themselves, there is often a moment of realisation; they realise they're not as compassionate or supportive to themselves as they would be to others in the same situation. Moving forward, the task would be for the students to practise giving themselves the advice they would ordinarily give to others. The concept of self-compassion is particularly relevant in the context of Maths Anxiety, given what we know about the negative self-belief and shame which are associated with feeling maths anxious. This is an interesting area for future empirical work, with lots of potential.

This method of appraisal may be beneficial within the classroom, particularly in ways which promote collaboration and facilitate the discussion of emotions related to Maths. This might be best achieved one-on-one if a student is highly anxious, so that they feel comfortable sharing their emotions with a trusted adult. However, through time and practise, this could also be adapted within group or whole class settings, to encourage discussion of reappraisal and how self-talk should be just as kind as the way we talk to others.

Mohammed's Experience

Mohammed was troubled by the way he physically reacted in Maths class when he got stuck on Maths problems. Then he remembered what his teacher had said about the way a person's body changes when they are faced with a challenge. This made him rethink the sweaty palms he was experiencing every time he found Maths tricky. He also remembered that his palms feel sweaty when he's a bit nervous before a football match, but he usually puts that down to excitement and wanting to perform well. In that situation, he told himself, he doesn't really see it as a problem. From then on, he became less and less troubled by his hands feeling sweaty in Maths lessons.

This strategy may again be easier to develop within secondary school and further education settings, but primary school students shouldn't be overlooked. It is often found that primary school students are more aware of their emotions than they are given credit for. Therefore, using reappraisal as a strategy with younger children is also worthwhile but may need to be adapted in terms of questions and vocabulary used to ensure it is accessible to all.

Mathematical Mindset

As we identified earlier in Chapter 4, the mindset that an individual has can impact their engagement and relationship with Maths and feelings of Maths Anxiety. With a fixed mindset, believing that you are born with your ability and that no amount of effort can change it, progress in mathematical attainment can be halted, and negative beliefs and attitudes become apparent, alongside negative affect such as anxiety. For example, a fixed mindset might be seen when someone doesn't involve themself in the group Maths task because they have preempted failure, as they believe it's too challenging for them to understand.

Having a growth mindset, where you believe that effort can have a positive effect on outcomes, has been seen to limit these negative effects, so it is important to teach this to students and consistently model a growth mindset in Maths lessons. A growth mindset might be seen through a student viewing mistakes as a positive in Maths tasks, for example, and seeing them as an opportunity to learn and grow.

Based on findings that suggest highly maths anxious individuals are more likely to have a fixed mindset, it could follow that addressing their mindset might have some positive impact on their anxiety. There is a small amount of research evidence to suggest this is the case. While more is needed, this is a promising start.

> Samuel and Warner (2021) devised a Maths Anxiety intervention that combined mindfulness strategies, such as controlled breathing, with strategies based on mindset. The latter involved learning about the principles of the "Growth Mindset Theory," but also repeating positive statements concerning oneself, such as *"I am capable of understanding Maths."* The Maths teacher would also reframe students' statements that represented a fixed mindset into statements that represented a growth mindset. For example, if a student said, *"I don't know why I keep making the same mistake for every problem,"* the teacher would respond with, *"Take a look at the strategy that you used for each problem – did you use the appropriate formula? It's okay to make mistakes now so you can learn from them."* The teacher was also active in reinforcing the importance of effort, endurance, and openness to feedback, while also frequently offering verbal praise in order to create a culture of positivity, effort, and endurance. The intervention led to a statistically significant reduction in Maths Anxiety, with the reduction also representing a moderate-to-large effect in terms of effect size.

There are a few ways that teachers can support the shift from a fixed mindset to a growth mindset, and therefore, potentially, have a positive impact on Maths Anxiety. Alongside consistently modelling a growth mindset of their own, the language used can be of significance.

The "power of yet" is often used within classrooms, which promotes the importance of having a growth mindset. Reframing negative, fixed mindset statements from *"I can't do it,"* to *"I can't do it yet"'* uses language associated with a growth mindset and demonstrates how learning is a process. Discussing mistakes as opportunities, rather than failure, is another way in which to support this, as well as viewing the brain as a muscle, which is strengthened by the process of learning and making mistakes. All of this can be verbal, or supported by visual aids around the classroom, such as posters, for example.

Developing a growth mindset can also happen from one-on-one discussions or interventions, or occur within a small group setting. It doesn't have to be a whole class strategy, but it would be useful to immerse maths anxious individuals within a classroom culture that focuses upon, and celebrates, the use of a growth mindset. If a maths anxious individual is not engaging with this within the whole class setting, it may be useful to have one-on-one discussions with them, to see how they can be supported further in developing a growth mindset.

> Clark (2021) used growth mindset lessons as a daily enrichment period for students in the fourth grade in the US (9-10 years old). Findings suggest that these growth mindset lessons had a positive impact on students developing a growth mindset and reducing levels of Maths Anxiety.

Sharing these thought processes with students, discussing the importance of having a growth mindset and the benefits of this, and positively reinforcing behaviours or language which demonstrate a growth mindset can all go towards instilling this within the classroom culture, and the students. Time might be dedicated to these discussions each morning, or through a regular session each week, or alternatively through more informal conversations throughout Maths lessons. How and when this is done, is completely dependent on the teacher, and the class needs, including any time constraints. This is a strategy which will not negatively impact money or resources, but it does require consistency when promoting the importance of having a growth mindset, so that students are engaged with the concept and take it on board. This may impact time initially, but when it is part of the classroom culture, this should no longer be time-consuming as it will be part of daily conversations if successfully embedded.

Mathematical Resilience

A comparatively broad approach, developing Mathematical Resilience aims to reduce Maths Anxiety by encouraging individuals to reflect on which situations are anxiety-evoking to them and work towards developing resilience in these situations so that they do not become all-consuming and negatively impact their well-being.

> According to Johnston-Wilder et al. (2020), being mathematically resilient means a person is not just confident and highly satisfied in the context of Maths, but they have a sense of value, purpose, or meaning. Further to this, they suggest that a mathematically resilient person is able to overcome threats to their self-efficacy when it comes to Maths.

A student who hasn't yet developed Mathematical Resilience may come across a challenging mathematical situation, such as feeling panicked when attempting a Maths problem they perceive as tricky, and not persevere. They might feel that they can't achieve and that their anxiety is overwhelming them. Instead, a student with Mathematical Resilience will tackle the challenging situation differently and feel equipped to overcome their feelings of anxiety and any setbacks that they may face.

Mathematical Resilience is not simply about surviving Maths situations where a person feels stressed or anxious. It is incumbent upon educators to ensure each and every student thrives and receives an enjoyable, satisfying, and useful Maths education. As such, this takes a combined approach to tackling Maths Anxiety and other psychological and contextual barriers.

> Lee and Johnston-Wilder (2017) outlined four key aspects needed for an individual to develop Mathematical Resilience:
>
> 1. Having a growth mindset
> 2. Knowing that Maths can have a personal value
> 3. Knowing how to work at Maths (Growth Zone Model)
> 4. Knowing how to find support

The first main factor within the development of Mathematical Resilience is ensuring that the individual has a *growth mindset*, rather than a fixed mindset, which we have previously discussed.

Students also need to know that *Maths can have a personal value*. This refers to the importance of teaching students why the mathematical content they are learning is useful and applicable in real-life situations. For example, during a lesson focusing on percentages, a teacher may lead the discussion of why this is useful when shopping, for example, or when calculating taxes or budgets. Students need to understand how they will use the mathematical concepts in their everyday lives in order to understand the relevance of being mathematically resilient, and not allow their anxiety to overtake their response to learning and using Maths. To do this, teachers may plan in time within lessons to discuss the utility of the lesson in everyday situations. Alternatively, it may be beneficial to have these discussions one-on-one with maths anxious individuals, as each student has an individual background, with different interests and needs. Therefore, by thoroughly understanding how they spend their time outside of school, for example, consideration can be given to how Maths is relevant to their context, and their life, with more specific examples.

The third factor, *knowing how to work at Maths*, is quite challenging to develop in maths anxious individuals. Having previous experiences which elicit bad memories, and associated anxiety and negative affect, can hinder how a student approaches different mathematical situations. As we have discussed before, it can be quite challenging to reframe these negative beliefs and identify the emotions, and emotional responses, that a maths anxious individual may be feeling. Here, the "Growth Zone Model" proves useful in helping students articulate how they are feeling about the mathematical situation they are in, and whether they need to take action to improve their current emotions.

To do this, the Growth Zone Model asks individuals which zone they feel that they are in:

THE COMFORT ZONE: where students are learning and using Maths they have done previously and feel confident in doing so, yet doesn't result in progression of learning and can lead to disengagement and boredom.
THE GROWTH ZONE: where students feel slightly challenged but in a positive way, leading to progression and growth in their understanding, without negative affect.
THE ANXIETY ZONE: where students feel fear, anxiety and panic, which overwhelms their capabilities to learn, stops learning progression, and can negatively impact their wellbeing.

Students clearly need to avoid being in the Anxiety Zone, but the Comfort Zone is not the place for successful, resilient students to be either. Therefore, the aim is for students to firstly identify where they feel they are within the zones, and if they are not in their personal growth zone, take action to move themself into this zone instead.

> Chisholm (2017) conducted interventions to develop Mathematical Resilience in Key Stage 4. Afterwards, participants had a better understanding of their experience of Maths Anxiety, were less avoidant, and were more aware of their next steps to improve their understanding of Maths and the skills of resilience.

To utilise this successfully will require detailed discussions with the maths anxious individual, not only of what the zones represent and how their emotions can represent these zones, but also, what actions can actually be taken to help them. This is most likely to be effectively used one-on-one, to really elicit an individual's understanding of their emotions and provide contextualised, meaningful actions that are relevant to their experience. This may require regular sessions initially, with sessions tailing off in frequency once the individual feels more confident to utilise this approach with more independence. However, having the Growth Zone Model displayed in the classroom, and referred to within whole class discussions, therefore embedding discussion and understanding of emotions with the classroom culture, can also be beneficial in reducing anxiety.

> Kirkland (2020) adopted a Mathematical Resilience intervention in three case studies focusing on Maths Anxiety in primary school. Using the Growth Zone model proved useful for the students in identifying and understanding their emotions when using Maths, and all students shared that they understood the model and felt that it had a positive impact on their feelings of anxiety.

Finally, *knowing how to find support* is the fourth essential tool in developing Mathematical Resilience. Maths anxious individuals need to be aware of what to do if they are feeling anxious and need support. This support may be related to not understanding the lesson content, or it may be that they need support in which resources to use in the classroom to help them approach a Maths question. Alternatively, it could be that they need support in regulating their emotions and intrusive thoughts. Regardless of the type of help they need, maths anxious individuals need to be taught how to find the support they need in order to reduce their feelings of anxiety. Within the classroom, this might be in the form of peer discussion, or adult support, or knowing how to utilise visual aids or manipulatives, while at home this might be asking for help from a family member. This does need to be a taught skill though, particularly with younger students, as they may be unlikely to have sought support before and may not have the know-how regarding what support to find, or how.

> **Latisha's Experience**
>
> Latisha's Maths Anxiety had grown so high that she felt she simply couldn't engage any more with Maths. Her teacher explained to her the concept of the Growth Zone Model and that it's normal for people to feel different degrees of calm or anxiety in response to different Maths problems. Latisha was encouraged to colour code each Maths problem that she worked on, using a different colour to represent the zone that she believed best reflected how she felt about each problem. After doing this for a few days, Latisha's teacher emphasised just how colourful her work was and how this demonstrated the range of thoughts and feelings she must have about the different Maths problems. This helped Latisha work towards tackling her fixed mindset. After several weeks of carrying on with the activity, Latisha was able to see how her colour coding had changed, demonstrating how she was now coding some Maths problems as being in the Growth Zone when previously, similar problems had been in the Anxiety Zone.

Mathematical Resilience, then, goes beyond targeting only Maths Anxiety, but addressing Maths Anxiety is still at its core. It acknowledges the multifaceted nature of Maths Anxiety and how it is interrelated with a range of emotions, attitudes, and the wider context in which people engage with Maths. The work of Sue Johnston-Wilder and various colleagues has been instrumental in this regard.

> Johnston-Wilder et al. (2020) published a toolkit for teachers to use when supporting Maths students to manage their emotions, and a full description is offered in their paper. In brief, this includes three tools:
>
> **THE HAND MODEL OF THE BRAIN:** Developed by Siegel (2010), this enables students to move their hand and fingers in a way that depicts the brain's fight or flight response. It offers a useful way for people to understand why and how anxiety and panic might occur.

> **THE RELAXATION RESPONSE**: Based on the work of Benson (2000), this includes a range of breathing and micro-mindfulness activities to self-soothe by engaging the parasympathetic nervous system.
>
> **THE GROWTH ZONE MODEL**: This provides a framework for students to distinguish between perceived challenge and threat in the context of Maths, enabling them to state and communicate their current feelings.

The toolkit produced by Johnston-Wilder and colleagues, and subsequently extended and modified by others, has gained widespread recognition, with plenty of examples demonstrating its efficacy in addressing Maths Anxiety. We are unable to cover this in sufficient depth here to do it justice, but we urge readers to explore the materials available from researchers in this area.

> Johnston-Wilder and Lee's (2024) *The Mathematical Resilience Book: How Everyone Can Progress in Mathematics* is an excellent source from which to learn more about Mathematical Resilience, including discussion of how teaching for Mathematical Resilience can help mitigate Maths Anxiety.

So... What Can We Do?

- To be effective, strategies should be tailored to the individual's experience of Maths Anxiety, so it is important that this is understood first.
- Improving Maths skills and understanding can have a positive impact on Maths Anxiety, but this would be more effective when teamed with other strategies addressing the underlying anxiety.
- Teachers can have a positive impact on Maths Anxiety through their language, behaviour, and task choices, as well as through pedagogical approaches such as using stories, cooperative learning, and flipped learning.
- Relaxation techniques, including focused breathing utilise mindfulness to overcome anxiety.
- Systematic desensitisation is an effective behavioural intervention, but requires specialist training.
- Strategies designed to express one's Maths Anxiety in a creative way might offer several possibilities in reducing it.
- Cognitive restructuring and reappraisal are effective cognitive strategies to reduce Maths Anxiety, as is increasing the individual's self-awareness of their emotions and avoidant behaviour. Peer coaching may assist in this.
- Developing mathematical mindsets and Mathematical Resilience provides maths anxious individuals with the tools to identify and overcome their anxiety.
- Engaging parents in the strategies chosen to support maths anxious students can lead to more effective outcomes, due to consistency and alignment of the home-school relationship.

Here, we have discussed some of the targeted interventions for Maths Anxiety that have been published in the academic literature. There are several advantages to the rigorous approach to testing these, along with conclusions based on appropriate statistical testing or systematic qualitative analysis. However, we should always bear in mind the complex reality in which Maths Anxiety exists. In particular, as a teacher or other professional working in Maths education, it is always important to consider both the individual student experience, but also the wider context in which the student experiences Maths. In addition to this, professional training, resources, students' general attitudes, and co-occurring psychological barriers all need to be factored into the selection of the most appropriate strategies.

As you will have seen, some strategies are based largely on cognition and the way in which students approach Maths, including dealing with unhelpful thoughts or beliefs. Often, a good starting point is to support students in self-identifying the kinds of self-defeating thoughts they might be having. If they understand the negative impact of these thoughts, it can help them to then understand that lower or decreased Maths performance is not due to their lack of ability. Other strategies are based mostly on students' behaviour, often acknowledging and tackling avoidance. A common thread throughout various intervention studies is the need to regulate emotions in order to overcome Maths Anxiety, so strategies such as relaxed breathing and mindfulness might have an important part to play.

It is likely that a combined approach offers the best solution to solving the equation of Maths Anxiety. Students should be included as active participants, including regular dialogue to ensure they are listened to, so that interventions can be monitored and adjusted where necessary. Teachers and school Maths leaders should feel empowered to take the reins in making Maths Anxiety a strategic priority, using evidence-based approaches while having creative and professional freedom to ensure that strategies meet the needs of their students.

As a final note, while we have touched upon the link between schools, students, and parents when considering how to address Maths Anxiety, this is an area that requires much more attention. School Maths leaders might want to consider if it is helpful to disseminate to parents how their school is tackling Maths Anxiety, and if so, what this might look like. It should also be acknowledged that parents themselves might benefit from some of the strategies discussed, which in turn might be beneficial for parent-child Maths interactions, especially when it comes to Maths homework. Workshops, talks, webinars, and other ways of generating conversations about Maths Anxiety with parents can be powerful, facilitating a more compassionate and understanding approach to tackling Maths Anxiety all round.

References

Akeb-Urai, N., Kadir, N. B. Y. A. and Nasir, R. (2020) 'Mathematics Anxiety and Performance Among College Students: Effectiveness of Systematic Desensitization Treatment', *Intellectual Discourse*, 28(1), pp. 99-127.

Arch, J. J. and Craske, M. G. (2006) 'Mechanisms of Mindfulness: Emotion Regulation Following a Focused Breathing Induction', *Behaviour Research and Therapy*, 44, pp. 1849-1858. Available at: https://doi.org/10.1016/j.brat.2005.12.007.

Asikhia, O. A. (2014) 'Effect of Cognitive restructuring on the Reduction of Mathematics Anxiety Among Senior Secondary School Students in Ogun State, Nigeria', *International Journal of Education and Research*, 2(2), pp. 1-20.

Balt, M., Bornert-Ringleb, M. and Orbach, L. (2022) 'Reducing Math Anxiety in School Children: A Systematic Review of Intervention Research', *Frontiers in Education*, 7, pp. 1-15. Available at: https://doi.org/10.3389/feduc.2022.798516.

Benson, H. (2000) The Relaxation Response. New York: Avon Books. Available at: https://doi.org/10.1080/00332747.1974.11023785.

Bishop, J. and Verleger, M. A. (2013) 'The Flipped Classroom: A Survey of the Research', 2013 ASEE Annual Conference & Exposition, 23, pp. 1-18. Available at: https://dx.doi.org/10.18260/1-2--22585.

Brunyé, T. T., Mahoney, C. R., Giles, G. E., Rapp, D. N., Taylor, H. A. and Kanarek, R. B. (2013) 'Learning to Relax: Evaluating Four Brief Interventions for Overcoming the Negative Emotions Accompanying Math Anxiety', *Learning and Individual Differences*, 27, pp. 1-7. Available at: https://doi.org/10.1016/j.lindif.2013.06.008.

Chisholm, C. (2017) The Development of Mathematical Resilience in KS4 Learners. Ph. D. Thesis. University of Warwick.

Clark, J. E. (2021) Case Study of Growth Mindset and the Impact on Math Anxiety. Ph. D. Thesis. Brenau University.

Codding, R. S., Goodridge, A. E., Hill, E., Kromminga, K. R., Chehayeb, R., Volpe, R. J. and Scheman, N. (2023) 'Meta-Analysis of Skill-Based and Therapeutic Interventions to Address Math Anxiety', *Journal of School Psychology*, 100, p. 101229. Available at: https://doi.org/10.1016/j.jsp.2023.101229.

Cropp, I. (2017) 'Using Peer Mentoring to Reduce Mathematical Anxiety', *Research Papers in Education*, 32(4), pp. 481-500. Available at: https://doi.org/10.1080/02671522.2017.1318808.

Davis, T. and Kahn, S. (2018) 'Management of Mathematics Anxiety: Virtual Relaxation Interventions', *International Journal of Education and Social Science*, 5(7), pp. 38-44.

Dorothea, G. (2016) 'Effectiveness of Psychodrama Group Therapy on Pupils with Mathematics Anxiety', *Zeitschrift für Psychodrama und Soziometrie*, 15(S1), pp. 197-215. Available at: https://dx.doi.org/10.1007/s11620-015-0299-4.

Gan, S. K. E., Lim, K. M. J. and Haw, Y. X. (2016) 'The Relaxation Effects of Stimulative and Sedative Music on Mathematics Anxiety: A Perception to Physiology Model', *Psychology of Music*, 44(4), pp. 730-741. Available at: https://doi.org/10.1177/0305735615590430.

Ganley, C. M., Conlon, R. A., McGraw, A. L., Barroso, C. and Geer, E. A. (2021) 'The Effect of Brief Anxiety Interventions on Reported Anxiety and Math Test Performance', *Journal of Numerical Cognition*, 7(1), pp. 4-19. Available at: https://doi.org/10.5964/jnc.6065.

Hembree, R. (1990) 'The Nature, Effects, and Relief of Mathematics Anxiety', *Journal for Research in Mathematics Education*, 21(1), pp. 33-46. Available at: https://doi.org/10.2307/749455.

Hines, C. L., Brown, N. W. and Myran, S. (2016) 'The Effects of Expressive Writing on General and Mathematics Anxiety For a Sample of High School Students', *Education*, 137(1), pp. 39-45.

Hunt, T. E. and Maloney, E. A. (2022) 'Appraisals of Previous Maths Experiences Play an Important Role in Maths Anxiety', *Annals of the New York Academy of Sciences*, 15(1), pp. 143-154. Available at: https://doi.org/10.1111/nyas.14805.

Jamieson, J. P., Peters, B. J., Greenwood, E. J. and Altose, A. J. (2016) 'Reappraising Stress Arousal Improves Performance and Reduces Evaluation Anxiety in Classroom Exam Situations', *Social Psychological and Personality Science*, 7(6), pp. 579-587. Available at: https://doi.org/10.1177/1948550616644656.

Johnston-Wilder, S., Baker, J. K., McCracken, A., and Msimanga, A. (2020) 'A Toolkit for Teachers and Learners, Parents, Carers and Support Staff: Improving Mathematical Safeguarding and Building Resilience to Increase Effectiveness of Teaching and Learning Mathematics', *Creative Education*, 11, pp. 1418-1441. Available at: https://doi.org/10.4236/ce.2020.118104.

Johnston-Wilder, S., and Lee, C. (2024) The Mathematical Resilience Book: How Everyone Can Progress in Mathematics. London: Routledge. Available at: https://doi.org/10.4324/9781003334354.

Johnston-Wilder, S. and Marshall, E. (2017) 'Overcoming Affective Barriers to Mathematical Learning in Practice', in The Open University (ed.) Proceedings of the Second International Conference on Developing Mathematical Resilience. Milton Keynes: Mathematical Resilience Network.

Kirkland, H. R. (2020) An Exploration of Maths Anxiety and Interventions in the Primary Classroom. Ph. D. Thesis. University of Leicester.

Kirkpatrick, A. (2020) A Pilot Study: What is The Impact of Daily Sessions with a Therapy Dog on the Attitudes of Anxious Pupils Towards School. Available at: https://my.chartered.college/impact_article/a-pilot-study-what-is-the-impact-of-daily-sessions-with-a-therapy-dog-on-the-attitudes-of-anxious-pupils-towards-school/ (Accessed: 12 May 2024).

Lavasani, M. G. and Khandan, F. (2011) 'The Effect of Cooperative Learning on Mathematics Anxiety and Help Seeking Behavior', *Procedia-Social and Behavioral Sciences*, 15, pp. 271–276. Available at: https://dx.doi.org/10.1016/j.sbspro.2011.03.085.

Lee, C. and Johnston-Wilder, S. (2017) 'The Construct of Mathematical Resilience', in Eligio, U. X. (ed.) *Understanding Emotions in Mathematical Thinking and Learning*. Massachusetts: Academic Press, pp. 269–291.

Mehdizadeh, S., Nojabaee, S. and Asgari, M. H. (2013) 'The Effect of Cooperative Learning on Math Anxiety, Help Seeking Behavior', *Journal of Basic and Applied Science Research*, 3(1), pp. 1185–1190. Available at: https://dx.doi.org/10.1016/j.sbspro.2011.03.085.

Meichenbaum, D. (1977) *Cognitive Behaviour Modification: An Integrative Approach*. New York: Plenum Press.

Moliner, L. and Alegre, F. (2020) 'Peer Tutoring Effects on Students' Mathematics Anxiety: A Middle School Experience', *Frontiers in Psychology*, 11, p. 527714. Available at: https://doi.org/10.3389/fpsyg.2020.01610.

Niaei, S., Imanzadeh, A. and Vahedi, S. (2021) 'The Effectiveness of Flipped Teaching on Math Anxiety and Math Performance in 5th Grade Students', *Technology of Education Journal*, 15(3), pp. 419–428. Available at: https://doi.org/10.22061/tej.2020.5908.2303.

Núñez-Peña, M. I., Bono, R. and Suárez-Pellicioni, M. (2015) 'Feedback on Students' Performance: A Possible Way of Reducing the Negative Effect of Math Anxiety in Higher Education', *International Journal of Educational Research*, 70, pp. 80–87. Available at: https://doi.org/10.1016/j.ijer.2015.02.005.

Pantino, F. O. and Hondrade-Pantino, J. (2021) 'Effects of Creative Writing Activities on Students' Mathematics Anxiety', *Journal of International Education*, 3, pp. 22–42.

Parameswara, A. A., Utami, S. W. and Eva, N. (2022) 'The Effectiveness of Cognitive Restructuring Techniques to Reduce Mathematics Anxiety in High School Students', *Journal for the Mathematics Education and Teaching Practices*, 3(1), pp. 31–43.

Park, D., Ramirez, G. and Beilock, S. L. (2014) 'The Role of Expressive Writing in Math Anxiety', *Journal of Experimental Psychology: Applied*, 20(2), p. 103. Available at: https://doi.org/10.1037/xap0000013.

Passolunghi, M. C., De Vita, C. and Pellizzoni, S. (2020) 'Math Anxiety and Math Achievement: The Effects of Emotional and Math Strategy Training', *Developmental Science*, 23(6), p. 12964. Available at: https://doi.org/10.1111/desc.12964.

Petronzi, D., Hunt, T. E. and Sheffield, D. (2021) 'Interventions to Address Mathematics Anxiety: An Overview and Recommendations', in Kiray, S. A. and Tomevska-Ilievska, E. (eds.) *Current Studies in Educational Disciplines*. Konya: ISRES Publishing, pp. 169–194. Available at: https://dx.doi.org/10.31234/osf.io/a46eh.

Petronzi, D., Schalkwyk, G. and Petronzi, R. (2024) 'A Pilot Math Anxiety Storybook Approach to Normalize Math Talk in Children and to Support Emotion Regulation', *Journal of Research in Childhood Education*, 38(1), pp. 145–163. Available at: https://doi.org/10.1080/02568543.2023.2214591.

Pizzie, R. G., McDermott, C. L., Salem, T. G. and Kraemer, D. J. (2020) 'Neural Evidence for Cognitive Reappraisal as a Strategy to Alleviate the Effects of Math Anxiety', *Social Cognitive and Affective Neuroscience*, 15(12), pp. 1271–1287. Available at: https://doi.org/10.1093%2Fscan%2Fnsaa161.

Ramirez, G. and Beilock, S. L. (2011) 'Writing About Testing Worries Boosts Exam Performance in the Classroom', *Science*, 331(6014), pp. 211–213. Available at: https://doi.org/10.1126/science.1199427.

Salazar, L. R. (2019) 'Exploring the Effect of Coloring Mandalas on Students' Math Anxiety in Business Statistics Courses', *Business, Management and Education*, 17(2), pp. 134–151. Available at: https://dx.doi.org/10.3846/bme.2019.11024.

Sammallahti, E., Finell, J., Jonsson, B. and Korhonen, J. (2023) 'A Meta-Analysis of Math Anxiety Interventions', *Journal of Numerical Cognition*, 9(2), pp. 346–362. Available at: https://dx.doi.org/10.5964/jnc.8401.

Samuel, T. S. and Warner, J. (2021) '"I Can Math!": Reducing Math Anxiety and Increasing Math Self-Efficacy Using a Mindfulness and Growth Mindset-Based Intervention in First-Year Students', *Community College Journal of Research and Practice*, 45(3), pp. 205–222. Available at: https://dx.doi.org/10.1080/10668926.2019.1666063.

Segumpan, L. L. B. and Tan, D. A. (2018) 'Mathematics Performance and Anxiety of Junior High School Students in a Flipped Classroom', *European Journal of Education Studies*, 4(12), pp. 1–33. Available at: https://dx.doi.org/10.5281/zenodo.1325918.

Sheffield, D. and Hunt, T. (2007) 'How Does Anxiety Influence Math Performance and What Can We Do About It?', *MSOR Connections*, 6(4), pp. 19–21.

Siegel, D. (2010) Mindsight: Transform Your Brain with the New Science of Kindness. London: Oneworld Publications.

Soll, B. and Petronzi, D. (2023) Validation of a Therapy Dog Intervention for Maths Anxiety. Derby: University of Derby Undergraduate Research Scholarship Scheme.

Supekar, K., Iuculano, T., Chen, L. and Menon, V. (2015) 'Remediation of Childhood Math Anxiety and Associated Neural Circuits Through Cognitive Tutoring', *Journal of Neuroscience*, 35(36), pp. 12574-12583. Available at: https://doi.org/10.1523/JNEUROSCI.0786-15.2015.

Vokatis, B. (2023) The Magic of Innovative Dog Therapy. Available at: https://medium.com/@bmvokatis/the-magic-of-innovative-dog-therapy-995b2979a514 (Accessed: 15 May 2024).

Walter, H. (2018) The Effect of Expressive Writing on Second-Grade Math Achievement and Math Anxiety. Ph. D. Thesis. George Fox University.

Willis, J. (2010) Learning to Love Math: Teaching Strategies That Change Student Attitudes and Get Results. Vancouver: ASCD.

7 Can Teachers Be Maths Anxious, Too?

To put it simply, yes, even teachers can experience Maths Anxiety.

Increasingly, schools have recognised that Maths Anxiety exists in their students and that a targeted approach is required to support the school with addressing the issue. Therefore, there have been more workshops, conferences, talks, and continuing professional development sessions which focus on Maths Anxiety than ever in the last few years. This increase in activity and focus has also led to conversations regarding teachers' own Maths Anxiety. It began with the occasional teacher tentatively mentioning:

"I used to be Maths anxious," or,
"I know teachers who are Maths anxious."

The increase in attention on teachers' anxiety and confidence with Maths has now led to some teachers coming forward to openly discuss how they feel anxious when it comes to Maths, at least in some contexts.

> Coppola et al. (2013) asked 90 trainee primary teachers to write three words they associate with the word "Mathematics." Over a quarter only wrote words that reflected negative emotions. Three quarters wrote at least one of the following words: fear, anxiety, stress, distress, tension, anger, anguish, affliction, dread, boredom, panic, discouragement, depression, repulsion, revulsion, frustration, or unease.

There has also been a wider acknowledgement of the personal challenges faced by teachers. National Numeracy, for example, has led discussion on this topic, providing a platform for teachers such as Gillian Lynch to tell others how they overcame Maths Anxiety. The Maths Anxiety Trust also has resources dedicated to supporting teachers who experience Maths Anxiety, including a brief guide that Tom co-authored with Sue Skyrme. In addition to this, conferences, such as those organised by the PiXL Network, have targeted leaders in Maths education, recognising how widespread and important the issues are. Now we are starting to hear Maths leaders openly ask:

"I have a lot of Maths anxious teachers in my school. What should I do?"
"How can I support teachers who are anxious about Maths?"

While many think that teachers are immune to the effects of Maths Anxiety, this is clearly not the case. We hear stories from people currently undertaking teacher training who tell us that the Maths component of their training, as well as expectations around teaching Maths, contribute to much of their anxiety. Some qualified teachers will openly admit that they do what they can to avoid Maths teaching, even after years in the job. Moreover, Tom recently heard from the leader of a teacher training programme for primary education that one of their students started the course without realising that they would be expected to even teach Maths.

> Mihalko (1978) suggested that teachers should not be expected to generate enthusiasm and excitement for a subject for which they have fear and anxiety.

There is an important distinction to make here though, and it often goes unidentified. A teacher may be experiencing Maths Anxiety, related to the Maths itself, or they may be experiencing anxiety about teaching Maths. Maths Teaching Anxiety is different from Maths Anxiety, which can be considered as being broader. While it is possible for a teacher to experience both, especially as they are overlapping concepts, it is important to consider and explore both concepts separately. This will allow us to better understand them, and the impact they can have upon a teacher, and the students they teach.

Teacher's Maths Anxiety

Empirical research into teachers experiencing Maths Anxiety themselves has only recently gained momentum, but literature on this has existed for some time. This has mainly focused on pre-service teachers, those who are in training, rather than with teachers who have been teaching for many years, with some exceptions. This is likely to impact findings, and it emphasises the need for further research into Maths Anxiety and more experienced teachers.

This leads us to ask whether certain teachers are more likely to be maths anxious than others. It may be possible that experienced teachers may feel less anxious than they initially did at the beginning of their teaching career due to repeated and regular exposure to Maths. The more they are around Maths, and required to engage with it, the less anxiety they may experience. However, it is also possible that the opposite may be true - that repeated exposure may in fact heighten a teacher's anxiety. Or perhaps repeated exposure has led experienced teachers who are maths anxious to develop coping strategies to utilise within the classroom, such as having the answers to all the questions written down within arm's reach, or having a trusty calculator to hand. This is clearly an interesting area to explore further.

> ### Luke's Experience
>
> Recalling how he used to hate Maths, he felt it was quite ironic that he was paid to teach it once a day to his class. Now in his third year of teaching, Luke felt really anxious when completing his teacher training, as he was asked to demonstrate his mathematical proficiency, of which he believed he had none. Luke shared that he thought

> he would fail, and that all the hard work and practice would have been for nothing, but he *"passed by the skin of his teeth."* He then began to worry about how he would teach Maths in Key Stage 2, but then felt lucky that he was teaching Year 1 (where he has remained since qualifying).

Related to the idea of exposure, specialist Maths teachers, such as those teaching solely Maths in secondary education and beyond, may be less likely to experience Maths Anxiety compared to non-specialist Maths teachers, as they are choosing to seek out opportunities to teach Maths, rather than avoid it. This is an important consideration when engaging with existing research, as a primary school teacher who teaches an hour of Maths a day, or an Art teacher who does not teach Maths explicitly at all, will both have different experiences to each other, let alone compared to a specialist Maths teacher. This can potentially influence their relationship with Maths, and their anxiety. This is not to say that a specialist Maths teacher has not, or cannot, experience Maths Anxiety, and that we should definitely not discount them from being potentially maths anxious, but it may be less likely.

> Hembree's (1990) meta-analysis showed that students preparing to be elementary (primary) school teachers reported the highest Maths Anxiety, compared with university students across different subject areas. This pattern was repeated in a later study by Baloglu and Kocak (2006).

The concept of maths anxious teachers is perhaps a sensitive topic though, for all sorts of reasons. In particular, highlighting that a teacher is maths anxious can create self-doubt in their own mathematical ability, but also their ability to teach the required knowledge to their students. While it is nothing to be ashamed of, this is likely to negatively impact their self-belief, and perhaps how they view themselves compared to other teachers who appear less maths anxious, or not maths anxious at all. Maths anxious teachers may be able to successfully utilise their experience though and use it to better understand and support maths anxious students in the classroom. Knowing how it felt, or feels, to be maths anxious may create a much deeper understanding of a student's experience of Maths Anxiety compared to a teacher who is not maths anxious, and never has been.

> Dove, Montague, and Hunt (2021) conducted an interpretative phenomenological study of four primary school teachers who self-declared themselves as being maths anxious. The researchers identified a theme regarding the consequences of experiencing Maths Anxiety as a teaching professional.
>
> Firstly, teachers who had experienced Maths Anxiety themselves felt more equipped to recognise Maths Anxiety in their students, including recognition of specific behaviours deemed to be associated with Maths Anxiety. For example, recognition of the

> consequences of either raising or not raising one's hand to answer a question posed by the teacher. From the student's perspective, putting forward an answer publicly in this way places them in a potentially vulnerable position if their confidence is knocked by providing an incorrect answer. Similarly, however, they may choose to avoid raising their hand, which opens up the risk that a teacher might single them out, coaxing them to provide a response. A quick, rambled, perhaps even nonsensical response could further indicate that the student is feeling anxious. Of course, such behaviours can be interpreted in different ways, but a teacher who empathises with Maths anxious students is more likely to see such behaviours as they experienced them when they were young students.
>
> The researchers further noted that teachers who themselves had experienced Maths Anxiety were more inclined to feel it helped their teaching. They felt they had more patience, could explain mathematical concepts better, and appreciated the need to be more flexible in their teaching.

We can also consider teachers and Maths Anxiety from a number of perspectives. As we touched on earlier in the book, students' perception of their Maths teacher's confidence and ability in Maths is important for their own Maths achievement and anxiety. Reducing and avoiding Maths Anxiety in teachers should therefore be prioritised from a student-centred point of view. As important as that is, Maths Anxiety can be stressful and tiring, so from a well-being perspective, a priority should also be to support those who teach Maths, whether that is supporting those who have coped with feeling maths anxious in the classroom for a number of years, or those who are only just embarking on their teacher training.

Underlying both of these areas of focus is the need to establish the multitude of reasons why teachers experience Maths Anxiety. Of course, teachers are human and what we know about the individual characteristics, self-beliefs, and early education and social experiences that underpin the development of Maths Anxiety will apply to teachers just as much as the general adult population.

> Bekdemir (2010) showed that pre-service teachers' Maths Anxiety was related to their own negative classroom experiences, in particular the extent to which they perceived their own teachers as inadequate and hostile.

Existing research highlights how teachers, including pre-service teachers, reflect on their own experiences of learning Maths in school when discussing what led to their Maths Anxiety developing. This may relate to having strict teachers and feeling unsupported, feeling that they were never good at Maths, or feeling embarrassed in front of the class. This is similar to the experiences of adults who are not teachers, highlighting that teachers have the same experiences growing up in the Maths classroom. This is a good time to remind ourselves that teachers were students once, often sharing experiences with fellow students who did not go on to choose a career in teaching. Findings such as these indicate that certain memories about Maths experiences in school can have a notable impact on someone's feelings towards the subject.

It is interesting that reports of negative school Maths experiences did not deter the people from embarking on a career as a teacher. Perhaps in many cases, the desire to teach outweighs the lasting effects of negative school experiences. Or, is it that some people demonstrate a certain degree of Mathematical Resilience that helps them push through their anxiety? It is feasible that some teachers who had a negative experience of Maths learning at school want to avoid creating those same experiences for their own Maths students, perhaps supporting a more empathetic approach to their own teaching practice.

> Uusimaki and Nason (2004) conducted interviews with pre-service teachers, which generated three "school experience" themes:
>
> - Origins of negative beliefs and anxiety about Maths;
> - Situations causing most Maths Anxiety; and
> - Types of Maths causing Maths Anxiety.
>
> They reported that most of the pre-service teachers' Maths Anxiety could be attributed to their own negative experiences of learning Maths at primary school. Such experiences involved test situations, poor teacher attitudes, having to give verbal explanations, and difficult mathematical content. The authors further reported that 72% of their participants attributed their negative school Maths experiences to their teachers.

However, as a group, teachers also experience scenarios and pressures that are specific to them. It is not our intent to politicise the discussion, but it is widely recognised that teachers are facing more and more pressure, with fewer resources available to schools and incremental changes to expected standards and targets in Maths. In particular, at primary level, some teachers report a mismatch between expected standards and their own Maths ability, leading them to feel overwhelmed and maths anxious. There is often a lot of discussion in primary schools, for example, where some qualified teachers feel that they cannot teach Year 6 because they can't understand the Maths that is required to be taught. After all, the taught mathematical content ranges so much from one academic year to the next, so why would someone who is maths anxious willingly agree to teaching a year group where the mathematical content may increase their anxiety and potentially lead to Maths Teaching Anxiety as well, if they didn't have to?

> Hunt et al. (2022) collected qualitative comments from trainee teachers in Romania, which led to interesting findings. Some individuals explained how they still experience Maths Anxiety but they mitigate this through increased effort, being organised, and regulating their emotions. Many wrote about positive and negative experiences with specific Maths teachers when they were in school themselves. For example, recalling negative experiences of learning Maths with teachers who they perceived as not being good at explaining mathematical concepts. Some also wrote about repeated changes in their Maths teacher and felt that having multiple teachers impacted consistency in their learning.

Maths Teaching Anxiety

It seems that the reality of teachers experiencing anxiety in relation to Maths has been somewhat swept under the carpet until now. Recent news articles have highlighted figures that suggest secondary schools in the UK are struggling to recruit Maths specialist teachers. This tallies with claims that many secondary schools are relying heavily on non-Maths specialists to teach Maths, and that those teachers are sometimes lacking in confidence and motivation to teach Maths. While a decline in recruitment may be for numerous reasons, one of these reasons may potentially relate to people's anxiety towards teaching Maths.

If a teacher is maths anxious, it stands to reason that they may also feel anxious about teaching Maths as well. While it is not always the case that both are experienced, there are definite similarities between the two concepts. Importantly though, we should not assume that a person who is generally anxious about Maths has an equivalent level of anxiety when it comes to actually teaching Maths. Likewise, anxiety associated with teaching Maths does not mean that that same teacher experiences anxiety when engaging in Maths tasks in other contexts. This distinction becomes especially important in the context of various pressures teachers might experience in relation to school and national targets for Maths.

For example, teachers often highlight the rapid pace at which they are required to work through the national curriculum for Maths. There are lots of objectives to teach the students each year, and if they don't, this could risk students falling behind and having gaps in their knowledge. If this happens every year, these gaps will only widen. This sense of urgency to teach concepts and move on to the next, sometimes irrespective of whether students have grasped the concept, can feed into Maths Teaching Anxiety.

> **Anna's Experience**
>
> Anna had been teaching in primary schools for over ten years but still felt the most anxious in Maths. She didn't feel anxious when teaching younger children, such as Year 2, but she felt that when she taught in Year 4 or higher (eight years old and more), she needed to review the content of each lesson a few times before teaching it. Anna felt that she didn't have the understanding needed to pass on to her class, especially as there were always some really capable students that needed challenging, but she didn't feel that she was able to.

Moreover, Year 6 teachers in England experience increased stress ahead of Standardised Assessment Tests for Maths, while secondary school Maths teachers have pressure from their GCSEs and A-Levels cohorts. If students don't achieve the grades they need, this can negatively impact their future education and career path. The outcome is of course reliant on the student's effort and motivation to succeed as well, but some teachers do feel stressed about student outcomes each day, with overwhelming pressure building throughout the year as they inch towards the national assessments. With this pressure stemming from expectations on student progress and achievement, as well as ensuring that everything that's needed to be taught is taught, it isn't surprising that teachers can experience anxiety about teaching Maths.

> Ganley et al. (2019) reported that teachers in lower primary school classes (with students aged 4-7 years) had greater anxiety about teaching Maths compared to those in upper primary (with students aged 7-11 years). They studied 399 primary-level teachers and found that self-reported Maths Teaching Anxiety was related to:
>
> - Lower Maths knowledge for teaching;
> - Lack of Maths-specific teaching credentials;
> - More traditional beliefs associated with Maths teaching and learning (transmissionist; fixed instructional plan; facts first).

Alongside workplace pressure, Maths Teaching Anxiety may interlink with a teacher's lack of self-belief in their Maths ability to effectively teach Maths to the level required for their job. This could be in the primary school setting, or it could be in secondary school where non-specialist teachers are required to cover Maths lessons. Without having the belief in their capability to teach Maths, regardless of whether this is an actual or perceived lack of ability, teachers may feel that they are negatively affecting students' learning and progression, perpetuating feelings of anxiety even more. On top of this, teachers' knowledge of the relevance of their own confidence and perceived competence could itself create pressure to perform, ultimately creating or exacerbating anxiety about teaching Maths. Thus, a vicious cycle could arise.

> Hunt and Sari (2019) conducted a study of 127 primary school teachers and trainees in the UK, using the Mathematics Teaching Anxiety Scale to assess teachers' anxiety specifically about teaching Maths. The results revealed that there is much variation between teachers concerning how anxious they are about teaching Maths; many teachers have little or no anxiety and several teachers reported more than being anxious just sometimes. Also, teaching experience was associated with anxiety, with more experienced teachers reporting less anxiety about teaching Maths.
>
> This research also showed that the Mathematics Teaching Anxiety Scale has two subscales. The first subscale, called "Self-Directed Mathematics Teaching Anxiety," was shown to measure anxiety associated with the teacher's own self-belief about their Maths teaching practice and ability. The second subscale, "Pupil/Student-Directed Mathematics Teaching Anxiety," concerns anxiety associated with their students' Maths understanding and performance. This highlights the complexity of understanding and measuring teachers' anxiety about teaching Maths.

It is difficult to say whether more experience in teaching causes a reduction in Maths Teaching Anxiety, perhaps through increased confidence, or whether there has been a generational shift. Anecdotally, many teachers we speak to suggest they have become increasingly less bothered by external demands or pressures as they have gained experience. They may also feel more comfortable with knowing the curriculum, what to teach, and how to teach it effectively to avoid potential misconceptions in their students. Nevertheless, we should not discount experienced teachers from the pool of educators who may potentially experience Maths Teaching Anxiety, particularly as we know that anxiety isn't constant, and that certain situations may evoke more anxiety than others. For example, teaching Maths to a small group of students rather than a whole class may be overwhelming for some teachers, who feel anxious that the students may be more likely to note their lack of confidence. Another teacher may not feel highly anxious about teaching Maths at all, until they have to teach the concept of fractions, and then they might become panicked and overwhelmed. Again, it's important to understand the individual's experience.

> Hunt et al. (2022) found that almost 17% of a sample of teachers experienced anxiety about teaching Maths more than just "sometimes," with the majority experiencing it rarely-to-sometimes.

The Impact of Teacher's Maths Anxiety

Whether it is Maths Anxiety, or anxiety about teaching Maths, what effect does this have on students?

Much of what we have discussed so far is based on teachers' own reports of their Maths Anxiety, or confidence in teaching Maths. We should also touch on the (slightly sensitive) possibility that maths anxious teachers behave in particular ways, say things, or change how they teach, without necessarily having an awareness of doing so. Of course, maths anxious teachers would not aim to increase the anxiety experienced by their students, but research has shown that it can happen.

While we discussed the influence of teachers who were not maths anxious in Chapter 5, it is worth exploring if maths anxious teachers have a specific effect on their students. In other words, could a teacher's Maths Anxiety influence students' attitudes and beliefs towards Maths in some way, perhaps indirectly? Very few studies have investigated this idea. This is understandable given the sensitivities surrounding such debates, but also because of the practical challenges with this type of research.

> Lau et al. (2022) analysed extensive data from three previous studies, spanning several countries and many thousands of students. Focusing on student-level variables, they found that the strongest predictor of Maths Anxiety was how competent students perceived their Maths teacher to be. More Maths Anxiety was reported in those who reported less confidence in their teacher. Reinforcing this, they also found the greater the confidence reported by teachers themselves, the lower the Maths Anxiety reported by students.

One hindering belief that may be inadvertently passed to students is that people have a fixed mindset, and that effort does little to impact someone's outcome in Maths. This can be shared with simple, everyday comments that reiterate how people are *"just naturally good at Maths,"* while others find it hard. While this is a potentially innocent phrase to share with students, this fixed mindset may be detrimental to someone's attitudes towards, and relationship with, Maths.

> Mizala, Martínez, and Martínez (2015) gave over 200 trainee teachers descriptions of primary school students. The descriptions gave details of the students' behavioural difficulties and underachievement in Maths. Unbeknownst to the teachers, the descriptions systematically referred to the student in each description as a boy or a girl. Teachers were asked to rate the expected achievement of the student, which resulted in some interesting, and slightly worrying, findings. Teachers who were higher in Maths Anxiety were more likely to give lower achievement expectations. Also, overall, girls' expected achievement was lower than boys. Such findings might tell us something about the ways in which Maths Anxiety might influence how teachers perceive their students, but they also indicate that student characteristics such as gender are relevant when it comes to teachers' expectations too.

A further belief that might be detrimental to students, and negatively influence Maths Anxiety, is gender stereotypes. While it is highly unlikely that a teacher nowadays will say to the class that boys are better at Maths than girls, or girls are better at other things than Maths, subtle actions can have a large effect. For example, by only asking boys challenging questions within a Maths lesson, this has the potential to negatively influence girls' perceptions of themselves in relation to Maths. Gender-related expectations may also be inadvertently shared, for example, through expecting male students to complete more work within a lesson than their female peers, or offering support much more regularly to girls than boys. Even though these small actions may be rooted in the desire to help, they may have a lasting effect on students' anxiety and perceptions of Maths.

> Beilock et al. (2010) not only assessed Maths Anxiety in female early elementary (primary) school teachers but also measured Maths achievement in their students. Importantly, the study was carried out longitudinally, which is more powerful than a typical cross-sectional (snapshot) study. They found that there was no association between a teacher's Math Anxiety and her students' Maths achievement at the beginning of the school year. However, by the end of the school year, the more anxious teachers were about Maths, the more likely girls were to endorse the commonly held stereotype that "boys are good at Maths, and girls are good at reading." Girls' Maths achievement was also lower. The same could not be said for the boys in the study. Studies such as this one suggest there is a great deal of complexity involved when trying to understand what is going on.

So, Can Teachers Be Maths Anxious, Too?

- Teachers can experience Maths Anxiety, too.
- Some teachers may not be maths anxious but instead can experience Maths Teaching Anxiety, which is distinct from Maths Anxiety itself.
- It is possible for a teacher to experience both Maths Anxiety and Maths Teaching Anxiety.
- Teaching experience, self-belief, confidence, work pressure, and the exact requirements of what they need to teach can influence both a teacher's Maths Anxiety and Maths Teaching Anxiety.
- Teachers' Maths Anxiety can influence Maths Anxiety in students, with teachers' maths confidence, perceived competence, and maths gender beliefs potentially playing a role.
- Policymakers, schools, and Maths leaders should take note of the research findings in this area, taking action to support the well-being and professional development of teachers and those undertaking teacher training.

While we can easily say that teachers can be maths anxious, it is much more complex than that. A teacher's anxiety towards Maths can be learnt just like anyone else's, and may be shaped by their own experiences with Maths, and influenced by their beliefs and attitudes. Interestingly, this may be the reason behind some people wanting to teach Maths in the first place, to support maths anxious individuals where they felt that they had not received the required support themselves.

A teacher's anxiety towards teaching Maths is more specific. While Maths Anxiety and Maths Teaching Anxiety are distinct from one another, they do both involve teachers having a lack of self-belief and confidence in their ability to understand Maths, and teach effectively. Curriculum pressures, expectations of student outcomes, and the need to teach Maths successfully can all feed into a teacher's Maths Teaching Anxiety as well though, which has the potential to negatively impact teacher recruitment and retention. Although not purposefully, Maths Anxiety and Maths Teaching Anxiety may also lead to anxiety developing in students, alongside unhelpful beliefs surrounding Maths.

It seems that we need to consider Maths Anxiety in teachers and students, plus their attitudes and beliefs surrounding Maths, but also take into account potentially subtle changes over time. On the same note, there have not been enough studies to justify strong claims about the influence of teachers' Maths Anxiety on their students. It is evident that the discussion is just beginning, and while more is being done to support teachers who are maths anxious, or are anxious about teaching Maths, this needs to continue in order to break the cycle and prevent teachers from feeling negative towards Maths.

References

Baloglu, M. and Kocak, R. (2006) 'A Multivariate Investigation of the Differences in Mathematics Anxiety', *Personality and Individual Differences*, 40(7), pp. 1325-1335. Available at: https://doi.org/10.1016/j.paid.2005.10.009.

Beilock, S. L., Gunderson, E. A., Ramirez, G. and Levine, S. C. (2010) 'Female Teachers' Math Anxiety Affects Girls' Math Achievement', Proceedings of the National Academy of Sciences. New York: PNAS, pp. 1860-1863. Available at: https://doi.org/10.1073/pnas.0910967107.

Bekdemir, M. (2010) 'The Pre-Service Teachers' Mathematics Anxiety Related to Depth of Negative Experiences in Mathematics Classroom While They Were Students', *Educational Studies in Mathematics*, 75(3), pp. 311-328. Available at: https://dx.doi.org/10.1007/s10649-010-9260-7.

Coppola, C., Di Martino, P., Mollo, M., Pacelli, T. and Sabena, C. (2013) 'Pre-Service Primary Teachers' Emotions: The Math-Redemption Phenomenon', Proceedings of the 37th Conference of the International Group for the Psychology of Mathematics Education, 2, pp. 225-232.

Dove, J., Montague, J. and Hunt, T. E. (2021) 'An Exploration of Primary School Teachers' Maths Anxiety Using Interpretative Phenomenological Analysis', *International Online Journal of Primary Education (IOJPE)*, 10(1), pp. 32-49.

Ganley, C. M., Schoen, R. C., LaVenia, M. and Tazaz, A. M. (2019) 'The Construct Validation of the Math Anxiety Scale for Teachers', *Aera Open*, 5(1), pp. 1-16. Available at: https://doi.org/10.1177/2332858419839702.

Hembree, R. (1990) 'The Nature, Effects, and Relief of Mathematics Anxiety', *Journal for Research in Mathematics Education*, 21(1), pp. 33-46. Available at: https://doi.org/10.2307/749455.

Hunt, T. and Sari, M. H. (2019) 'An English Version of the Mathematics Teaching Anxiety Scale', *International Journal of Assessment Tools in Education*, 6(3), pp. 436-443. Available at: https://dx.doi.org/10.21449/ijate.615640.

Hunt, T. E., Napiorkowska, A., Popa, I-L. and Bagdasar, O. (2022) 'Mathematics Teaching Anxiety in Romanian Pre-service Elementary Teachers', Mathematical Cognition and Learning Society Conference. 1-3 June 2022. Belgium: MCLS.

Lau, N. T., Hawes, Z., Tremblay, P. and Ansari, D. (2022) 'Disentangling the Individual and Contextual Effects of Math Anxiety: A Global Perspective', *Proceedings of the National Academy of Sciences*, 119(7), pp. 1-11. Available at: https://doi.org/10.1073/pnas.2115855119.

Mihalko, J. C. (1978) 'The Answers to the Prophets of Doom: Mathematics Teacher Education', in D. B. Aichele (ed.) *Mathematics Teacher Education: Critical Issues and Trends*. Washington: National Education Association, pp. 36-41.

Mizala, A., Martínez, F. and Martínez, S. (2015) 'Pre-Service Elementary School Teachers' Expectations About Student Performance: How Their Beliefs are Affected by Their Mathematics Anxiety and Student's Gender', *Teaching and Teacher Education*, 50, pp. 70-78. Available at: https://doi.org/10.1016/j.tate.2015.04.006.

Uusimaki, L. and Nason, R. (2004) 'Causes Underlying Pre-Service Teachers' Negative Beliefs and Anxieties about Mathematics', Proceedings of the 28th Conference of the International Group for the Psychology of Mathematics Education, 4, pp. 369-376.

8 Can We Solve the Equation?

As we have hopefully shown throughout this book, Maths Anxiety is multifaceted. It varies greatly between people, time, and specific mathematical contexts. Over several decades of research, it is clear that some features of Maths Anxiety are more typical than others, which has enabled researchers and educators to devise relatively broad approaches to address the issue. We can say with some confidence that Maths Anxiety is an actual anxiety and it's specific to Maths; it can't be reduced to other psychological constructs, including other forms of anxiety.

To some extent, it is helpful that research has focused on specific aspects of Maths Anxiety, enabling exact research questions and hypotheses to be addressed. It also means that there are an increasing number of meta-analyses and literature reviews that highlight where consistencies and inconsistencies lie. However, more answers have uncovered more questions, and we are operating in a Maths education context that is increasingly understood, but increasingly complex. This is an exciting time for Maths Anxiety research, but also one that brings challenges. The field has expanded exponentially, which has meant a greater number of interdisciplinary collaborations to understand and tackle Maths Anxiety. Indeed, this is needed given the complex nature of it. It also means that Maths Anxiety research is growing in many directions and it is important that we take this into account if we want to fully understand it, especially if we want to successfully address it.

We have shown that Maths Anxiety is related to a wide range of psychological variables, largely concerning attitudes and beliefs associated with Maths. An awareness of these relations furthers our understanding of Maths Anxiety itself but also gives some insight into the potential causal directions at play. The field has seen a slight shift towards more applied Maths Anxiety research, perhaps because of collective increased confidence in existing knowledge of the general nature of Maths Anxiety. It is also the case that teachers and other professionals working in Maths education require real-world solutions to real-world issues. Consequently, more questions are being asked regarding the ways in which research findings translate into purposeful practice.

A specific note should be made on the relation between Maths attainment and Maths Anxiety, as it is heavily featured within existing literature. The evidence is fairly clear that higher Maths Anxiety is associated with overall lower Maths attainment. This association is not quite as big as people assume, but it is present, and it can vary according to several

DOI: 10.4324/9781003426547-9

factors, including performance on specific Maths tests. It is likely, for instance, not only that Maths Anxiety and working memory interact in various ways to impact performance according to the type of Maths problems being solved but also that Maths Anxiety can affect how long it takes someone to perform in Maths as well as how much effort they expend. Given that Maths Anxiety can mask a student's true Maths ability, we should also consider the possibility that it is limiting the potential of some students; their grades might not be poor, just not as high as they could be. A focus on final Maths grades sometimes obscures other issues, including the effect of Maths Anxiety on a student's level of stress, their enjoyment of Maths, their self-beliefs, and decisions around what they go on to study or choose for work. As much as Maths Anxiety is about achievement and success in Maths, with clear economic implications when we consider the bigger picture, it is more an issue of well-being, of feeling psychologically safe and secure with one's Maths education and use of Maths in everyday life.

It is also clear that Maths Anxiety can be seen in children within the first years of formal education, which means tackling this head on should be a priority, especially as Maths Anxiety appears to ramp up around the midpoint of secondary school. We should also be mindful that some individuals do not fit the typical Maths Anxiety profile. They might be high attainers, perhaps moderately motivated to learn Maths, and might not have a particular issue with class tests. Tied with the fact that some students might suffer in silence, it is imperative that teachers feel confident in identifying the signs of Maths Anxiety. They should feel able to distinguish Maths Anxiety from other specific forms of anxiety and recognise that Maths Anxiety does not necessarily equate to low motivation or negative attitudes towards the subject.

We still have some way to go in this regard, but fortunately, Maths Anxiety is measurable, and this can be done in a reliable and valid way. There are now several self-report tools available to support such testing, often designed with specific groups of people in mind, and sometimes in reference to a particular context. Self-report measures are likely to remain popular, and we anticipate that more of these will be devised, with specific populations in mind, in the coming years. The usefulness of these tools should not be underestimated, whether it is to gauge the general level of Maths Anxiety in a class, cohort, or school, or whether it is to facilitate the identification of specific students in need of support. As we have seen, self-report measures can be used effectively as part of the intervention process too, helping students develop greater self-awareness of how they feel when it comes to Maths.

Undoubtedly, teachers, along with parents, employers, and many others, hold the power to address Maths Anxiety. Addressing Maths Anxiety provides a more inclusive Maths education environment, and there may be certain groups who are at-risk or in particular need of support. For example, there is evidence to suggest those with developmental dyscalculia are more likely to report higher Maths Anxiety, or that genetics, socio-economic background and culture can influence who is likely to be maths anxious. Females have also been reported to be more likely to experience Maths Anxiety than males. While this is not always the case, even a small difference in Maths Anxiety might be all that is needed to alter course or career choices, so it's important to work to address this.

Insights into each person's unique journey in Maths education have been provided by studies that involve people reporting their own Maths Anxiety, including interviews and surveys. Individual accounts need to be considered in order to unpack the ways in which students appraise their previous Maths experiences. So, when we talk about addressing Maths Anxiety, it is important to identify exactly which components of it are most in need of addressing with any given individual, whether that is one or more of the emotional, cognitive, or behavioural components.

Successful interventions appear to be those that acknowledge the individual and combine strategies. In particular, effective strategies address unhelpful thoughts and beliefs and help people to understand their own reactions to Maths, while also addressing negative self-belief and building Mathematical Resilience. Perhaps underpinning many successful interventions for Maths Anxiety is the role of compassion. That is, ensuring students feel their Maths Anxiety is not only acknowledged, but that we, as educators, are willing to be proactive in remedying the situation–in terms of alleviating Maths Anxiety, but also in preventing it.

Certainly, strategies that focus on emotion regulation have an advantage in helping students manage their anxiety, particularly in Maths situations in which they feel stressed or pressured. It goes without saying that many emotion regulation techniques also have the advantage of being transferable skills, enabling students to regulate their emotions in other, non-Maths, contexts. For a deeper impact, changes to teaching and learning practices can be combined with psychological strategies to address how students think about Maths–both in terms of the subject and in regards to their self-belief. A toolkit approach can help achieve this, building self-awareness and Mathematical Resilience to reduce Maths Anxiety. In many cases, however, existing interventions require further testing, particularly in relation to how they might be generalised beyond those who have already taken part.

When working to support a maths anxious individual, we also need to bear in mind that Maths Anxiety rarely exists in isolation. It is intertwined with someone's Maths attitudes, beliefs, and sometimes even psychological trauma from previous, negative Maths experiences. While a targeted approach can be highly effective in addressing Maths Anxiety specifically, a holistic approach that takes into account the wider learning experience might be needed when aiming for long-term outcomes, such as general Maths engagement. Thus, multiple individual characteristics and different contexts need to be considered.

If you have gotten this far, you might be wondering where to begin in taking action. By beginning to explore Maths Anxiety, you have already taken an important step, as taking an interest in Maths Anxiety is essential for understanding, recognising, or measuring it. It has been our aim not only to give some insight into our own experiences of working with maths anxious students but also to provide relevant information that is research-informed. We hope this book has provided a platform for reflection, contemplation, and planning, so that you are best placed to consider the changes that might be needed in whichever Maths context you are educating and supporting students. This might include a broad education context, such as whole-school approaches, but it might also be identification and modification of strategies that work for specific groups, such as students with Developmental Dyscalculia, or those training in particular professions, such as Nursing.

It is also important to recognise that there are some relatively fixed challenges, often based on external demands and available resources, which create resistance and difficulty in making desired changes. However, whether you are focusing on your own Maths teaching, or taking a broader perspective as a Maths leader, it is our ambition that this book motivates you to think about, and facilitate, positive change. This might include small steps, like beginning conversations about Maths Anxiety within your workplace, or encouraging self-reflection and the use of self-report tools. Based on what we have discussed about the prevalence of Maths Anxiety among teachers and the impact this can have, such actions are just as relevant in the context of teacher training and support networks. There are of course wider policies that need to consider the importance of Maths Anxiety within the teaching profession, and research involving teachers themselves will continue to strengthen calls to address this.

There is, of course, a huge amount of empirical research into Maths Anxiety that we have not been able to cover in the space of this book. Conducting your own literature searches, in addition to exploring the various reviews we have highlighted, is recommended. We expect Maths Anxiety research to take some interesting turns in the next few years. As the field moves forward, it is likely that we will see studies that increase in size and scope, making use of big data and collaborations between the many research groups and networks that specialise in the study of Maths Anxiety. We also anticipate that there will be more studies into Maths Anxiety in less developed countries, shedding further light on potential antecedents and challenges.

It may be likely that experimental studies will take an increasingly focused approach and provide us with much-needed information with specificity. This can already be seen to be happening, as research has started to explore the role of specific attentional processes that are thought to interact with Maths Anxiety to explain cognitive performance on Maths tasks. Further to this, the role of intrusive thoughts still requires some attention, along with the relevance of certain psychological factors that have been shown to be related to Maths Anxiety but not studied in sufficient depth. The psychological construct of shame, for instance, appears to be something that many Maths anxious students experience, but it has not received the attention it deserves.

From an applied perspective, there is also much more that we can do to test the relative impact of small, but potentially important, changes to teaching and learning practices. For instance, how does the way in which instructions are presented or framed to students with different levels of Maths Anxiety impact their experience? And what is the effect of specific features of cooperative learning upon Maths Anxiety across educational phases? We also expect to see more intervention-based studies that take an integrated approach, modifying specific strategies in various ways. Regarding this latter point, the close collaboration between researchers, teachers, and students is often missing. Only limited work has reported on the involvement of all parties in designing, testing, and modifying Maths Anxiety interventions, and we hope this is something that becomes commonplace.

Evidently, Maths Anxiety is prevalent across the general population and across all phases of education. It is widespread, suggesting a broad, system-level issue. While debates continue regarding the general nature of Maths education, there is much we, as educators, can do. We urge you to help spread awareness of Maths Anxiety, so that everyone is as informed as possible. We also encourage you to create psychologically safe mathematical learning environments, where students' anxiety and fears are preempted and addressed, fostering Mathematical Resilience. This can include minor modification to teaching practices, or more structured interventions.

Action research and working in partnership with academics in the field is vital, especially in understanding and tackling Maths Anxiety from a developmental perspective. Maths Anxiety is already rising to the top of the agenda of various organisations associated with teaching, leading us to believe that positive change is round the corner. Importantly though, teachers also require a supportive environment in which they can grow in confidence to teach Maths and interact with the wider network of people involved in a student's Maths education, such as parents. Teachers are so influential when it comes to students' feelings, emotions, and attitudes towards Maths.

So, can we solve the Maths Anxiety equation?

The answer is yes. Compared to where we were a few decades ago, there has been an enormous shift in our knowledge and understanding about Maths Anxiety: what it is, what it isn't, and what we can, and should do, to address it. Hopefully, this book has strengthened your understanding of Maths Anxiety and will support your endeavours in making a positive change.

INDEX

Abbreviated Maths Anxiety Scale (AMAS) 47
Academic Emotions Questionnaire 42
Academic Motivation Scale 40
achievement 15, 24, 59, 77, 79, 97, 101, 192, 195, 197
Achievement Goal Questionnaire 40
addition 66, 67, 128-129, 131
age 21, 28-29, 32, 49, 96, 101-102, 111, 123, 143, 147, 151-152, 156
Ahmadi, S. 32
Ahmed, W. 75, 78, 88
Aiken, L. R. 8, 42
Akeb-Urai, N. 165
Alegre, F. 152
algebra 12, 48, 112, 128, 134, 136, 166
American Psychiatric Association 8
American Psychological Association 11
animal therapy 172-174
Ansari, D. 62
appraisal 40, 177-178
Appraisal of Previous Mathematics Experiences Scale 40
approach-avoidance 40
Arch, J. J. 161
Arslan, C. 13
Ashcraft, M. H. 11, 20, 66-67, 69, 84, 87, 117
Asikhia, O. A. 175
assessment 59-60, 123-124, 195
Atkinson, R. T. 35
attainment 11-12, 17, 25, 38, 41, 43, 58-65, 67-70, 75-76, 80-81, 87-89, 96, 99, 102, 108, 111, 113, 146-148, 179, 200
attitudes 1-2, 12, 14, 17, 20, 24, 26, 30, 35-36, 46, 50-51, 58, 63-64, 71-73, 75, 81, 88, 94-95, 97, 100, 102-106, 108, 111, 124, 137, 143, 146, 148, 158, 173, 179, 183, 185, 193, 196-198, 200-202, 204
Attitudes Towards Mathematics Inventory (short form) 42
Attentional Control Theory 70
attentional processes 70, 83, 86, 203
Aunola, K. 60, 80
avoidance 3, 8-9, 12, 17, 58, 73-74, 81-88, 140, 157, 164, 185

Baddeley, A. D. 65
Balgalmis, E. 50
Baloglu, M. 50, 191
Balt, M. B. 144
Bartley, S. R. 95
Becker, M. 96
Beilock, S. L. 7, 10, 123, 167, 170, 197
Bekdemir, M. 192
Berkowitz, T. 98
Betz, N. 20
Bhardwa, J. 37
Bisanz, J. 50
Bishop, J. 149
Bokhorst, C. C. 108
Bono, R. 156
Bornert-Ringleb, M. 144
brain activity 2, 7, 21, 35, 37-39, 54
breathing techniques 140, 142, 160-162, 165, 179, 184-185
Brindley, J. 21
Brown, N. W. 168
Brunyé, T. T. 161
Burns, M. 20
Butler, R. 84

Cacioppo, J. T. 104
Calvo, M. G. 68
Capraro, M. M. 45
Capraro, R. M. 45
Carey, E. 15, 62
Chapman, E. 42
Children's Anxiety in Math Scale (CAMS) 50-51
Children's Mathematics Anxiety Scale UK (CMAS-UK) 52
Chinn, S. 9, 20, 124
Chisholm, C. 182
Chiu, L. H. 50
Cipora, K. 47, 63-64
Clark, J. E. 180
Clark-Carter, D. 37, 47-48, 70
Codding, R. S. 144, 149
cognition 1, 5, 9, 17, 27, 38, 94, 110, 169, 185
Cognitive Intrusions Questionnaire 41
cognitive restructuring 3, 141-142, 166, 174-176, 184
comfort zone 12, 126, 133, 135-136, 149, 182
conceptual understanding 117, 127, 133, 135, 149
confidence 4, 29, 36, 46, 58, 71, 74, 94, 114-116, 124, 130, 147, 151-152, 154-155, 172, 189, 192, 194-196, 198
Conroy, D. E. 40
cooperative learning 3, 119, 140, 152-154, 184, 203
Coppola, C. 189
cortisol 36-37
countries 23-25, 30, 32, 43, 47, 51-52, 196, 203
Craske, M. G. 161
Cropp, I. 151
culture 21, 23, 25, 30-32, 43, 103, 111-113, 117, 121-122, 124, 154, 162, 179-180, 182, 201
curriculum 23-24, 48, 51, 62, 81, 106, 115, 118, 126, 128, 131, 147, 194, 196, 198

Daane, C. J. 116
Daucourt, M. C. 101
Davis, T. 163
debilitating anxiety model 62
decimal numbers 125-126
deficit theory 61-62
Demedts, F. 36
Dent, P. 21
Devine, A. 30
De Vita, C. 142, 172

DiStefano, M. 101
division 12, 53, 125, 128, 131-132
Dorothea, G. 171
Dove, J. 52, 191
Dowker, A. 51, 74
Dowson, M. 77
Dreger, R. M. 8, 42
Dweck, C. S. 115
dyscalculia 60-61, 135, 201-202

Eccius-Wellmann, C. 23
Eccles, J. S. 77
Edwards, R. W. 49
effect size 142, 145-146, 179
Elliot, A. J. 40
Elliott, L. 102
emotional contagion 104-105
engagement 1, 71-73, 106, 123, 130, 148, 152-153, 163, 171, 179, 202
enjoyment 1, 4, 12, 36, 40, 42, 71, 73, 79-81, 84, 94, 103, 108, 110, 114, 118, 136, 201
environment 27-28, 66, 70-71, 78, 93, 105, 109, 113, 115, 121, 125, 141, 156, 162, 164, 171-172, 201, 204
Eva, N. 175
exams 15-16, 28, 59, 75, 87, 99, 122, 168
expectations 24, 32-33, 59-60, 83, 96, 103, 110, 113, 120, 123, 130-131, 148, 154, 156, 190, 195, 197-198
expressive writing 3, 141, 166-171
extrinsic motivation 40, 79
eye movements 36-37, 55
Eysenck, M. W. 68, 70

Fan, X. 25
Faust, M. W. 11, 66
fear 7-9, 14, 20, 24, 29, 31, 37, 40, 46, 51, 59, 71, 75, 82-83, 87, 94, 104, 109, 113-114, 118-125, 146, 150, 153-155, 173-174, 182, 189-190, 203
feedback 75, 121, 156, 179, 187
Fennema, E. 46
Fennema-Sherman Mathematics Attitudes Scale (FSMAS) 46
Ferguson, R. D. 128
Ferreira, R. A. 26
Finlayson, M. 118
Fiore, G. 98
Firouzian, F. 47

fixed mindset 24, 75, 97-98, 140, 179-181, 183, 197
Fleck, D. E. 66
flipped learning 3, 140, 148-151, 184
Foley, A. E. 98
fractions 12-13, 115, 117, 126-128, 131, 133, 136, 175, 196
Freeston, M. H. 41
Freudenthaler, H. H. 80
Friso-van den Bos, I. 65
Fugelsang, J. A. 62

Gan, S. K. E. 163
Ganley, C. M. 21, 169, 177, 195
Garba, A. 106, 107
gender 21, 23, 25, 30-32, 46-47, 120, 152, 197-198
general anxiety 15, 17, 28, 31, 36, 38, 64, 124
genetics 23, 27-28, 32, 94-95, 99, 110-111, 201
geometry 12, 133-134, 136
Gest, S. D. 104
Gibson, D. J. 98
Gierl, M. J. 50
Goetz, T. 31
Good, C. 115
Gough, M. F. 8, 19
grouping 118-119
growth mindset 24, 40, 112, 179-181
Guzman, B. 26

Hadfield, O. 11
Hajar, M. S. N. 21
Hambleton, R. K. 25
Hannula, M. S. 76
Harper, N. W. 116
Hart, S. A. 21
Hatfield, E. 104
Haw, Y. X. 163
heart rate 9, 15, 36-37, 123
Helmane, I. 118, 131
helplessness 11-12
Hembree, R. 29, 73, 77, 81, 142, 144, 191
Henry, L. L. 50
Henson, R. K. 45
Hill, F. 15
Hines, C. L. 168
Hitch, G. J. 65
homework 12, 24, 58, 86, 98, 100-102, 106, 109-111, 149, 157, 185

Hondrade-Pantino, J. 153
Hopko, D. R. 47
Hunt, T. E. 37, 40, 47-48, 52, 70, 97, 127, 134, 141, 144, 165, 174, 191, 193, 195-196

Imanzadeh, A. 150
Ingram, N. 95
instrumental motivation 40
interviews 13, 35, 52-55, 111, 172, 193
intrinsic motivation 40, 79-80
intrusive thoughts 9, 41, 70, 167-168, 170, 174, 183, 203
Ismail, N. 111

Jackson, C. D. 120
Jain, S. 77
Jameson, M. M. 50, 96
Jamieson, J. P. 176-177
Johnston-Wilder, S. 21, 78, 161, 164, 180-181, 183-184
Jones, W. G. 21

Kadir, N. B. Y. A. 165
Kahn, S. 163
Karim, A. H. 21
Kazelskis, R. 47
Keshavarzi, A. 32
Khandan, F. 154
Kim, J. 105
Kirk, P. E. 67
Kirkland, H. R. 52-53, 130, 140, 182
Kirkpatrick, A. 173
Kocak, R. 191
Kooken, J. 40, 42
Krause, J. A. 84, 117
Krinzinger, H. 51
Kumari, S. 95

Lane, C. 26, 100
language 12, 22, 39, 43-45, 49-50, 55, 72, 81, 94-95, 97-98, 103-109, 111-114, 121, 127, 143, 156-157, 178-180, 184
Lau, N. T. 196
Lavasani, M. G. 154
Learnus 9
Lee, C. 78, 181, 184
Leffingwell, R. J. 120

Leskinen, E. 60, 80
Leung, S. O. 46
Levine, S. C. 98
Libertus, M. E. 102
likert scale 16, 45-47, 49-52, 118
Lim, K. M. J. 163
Lim, S. Y. 42
longitudinal 15, 59-62, 70, 73, 76-78, 87, 105, 134, 197
Looi, C. Y. 74
Lyons, I. M. 7

Maas, I. 109
Mackrell, K. 78
Maijala, H. 76
Malanchini, M. 28
Maloney, E. A. 40, 62, 102, 174
manipulatives 117, 156-157, 183
Marshall, E. 161, 164
Martinez, F. 197
Martinez, S. 197
mathematical mindset 141, 179-180, 184
mathematical performance 11, 24-26, 30-32, 36-38, 40, 50-51, 59, 62, 64-65, 67-69, 73, 77, 80, 82-83, 96, 98, 121-123, 129, 143, 148-151, 156, 161, 163, 165-167, 169-170, 174, 176-177, 185, 195, 201, 203
mathematical resilience 3, 40, 42, 78, 141, 180-184, 193, 202-203
Mathematical Resilience Scale 40
Mathematics Anxiety Questionnaire (MAQ) 46-47
Mathematics Anxiety Rating Scale (MARS) 45, 49-50
Mathematics Anxiety Research Group 9
Mathematics Anxiety Scale-UK (MAS-UK) 47-48
Mathematics Anxiety Scale for Children (MASC) 50
Mathematics Anxiety Survey (MAXS) 50
Mathematics Attitudes Scale 46
Mathematics Calculation Anxiety Scale (MCAS) 48, 134
Maths Anxiety Questionnaires (MAQ) 22, 39, 42-51
Maths Anxiety Rating Scale-Elementary 49
Maths Anxiety Trust 9, 20, 28, 31, 189
Maths norms 41
Maths self-concept 41, 71, 74, 77-78

Maths self-efficacy 41, 71, 180
Maths teaching anxiety 3, 190, 193-196, 198
Maths utility 52, 71, 99, 127, 148, 181
Mattarella-Micke, A. 36
McLean, J. 25
measurement 32, 36, 38-39, 42, 47, 52, 54, 102, 134-136
Meece, J. L. 46, 77
Mehdizadeh, S. 153
Meichenbaum, D. 165
Menon, V. 38
meta-analysis 29, 73-74, 77, 91, 101, 141-144, 149, 166, 176, 191, 200
Mihalko, J. C. 190
Mindful Maths 9
mindfulness 160-164, 179, 184-185
Mizala, A. 197
modelling 27, 95, 99, 107, 109, 114, 162, 179
Moliner, L. 152
money 25, 51, 99, 111, 134-135, 180
Montague, J. 52, 191
Moore, A. M. 20, 69
motivation 1, 3, 5, 9, 12, 17, 26, 40-42, 46, 58, 60-61, 71-74, 79-83, 88-89, 93, 106, 108, 110, 112, 122, 124, 143, 150, 166, 194, 201
multiplication 8, 66, 87, 103, 124-125, 128, 130-132
Murayama, K. 40
Myran, S. 168

Nabila, L. A. 75
Nasir, R. 165
Nason, R. 193
National Mathematics Advisory Panel 69
National Numeracy 9, 20, 30, 189
Neubauer, A. C. 80
newly qualified teacher (NQT) 1, 23, 116
Niaei, S. 150
Nordin, N. M. 79
Núñez-Peña, M. I. 156
Nurmi, J. E. 60, 80

observations 1, 4, 13, 35, 53-54, 102, 156
online learning 149-150
Orbach, L. 144
Organisation for Economic Co-Operation Development (OECD) 24-25, 30, 40-41, 108
O'Toole, T. 113

Pantino, F. O. 153
Parameswara, A. A. 175
parental occupation 25-26, 99-100
parents 1, 3, 9-10, 12, 25-27, 41, 45-46, 54, 63, 72, 94-103, 107-111, 136, 141, 148, 150, 157, 162, 164, 172, 176, 184-185, 201, 204
Park, D. 105, 167-168
Passolunghi, M. C. 142, 172
peer mentoring 3, 140, 151-15
peers 3, 5, 12, 28, 41, 44, 60, 66, 73, 84, 93, 103-109, 113, 119-121, 123, 136, 151-157, 159, 168, 175, 183-184, 197
Pehkonen, E. 76
Pekrun, R. 42, 62
Pellizzoni, S. 142, 172
percentages 127, 181
perceptions 22, 41, 54, 71, 77, 82, 93, 95, 100, 103, 105, 107-113, 116, 119, 122, 128, 168, 197
Performance Failure Appraisal Inventory 40
Perry, A. B. 20
perspiration 15, 36, 55
Petronzi, D. 29, 52, 120-121, 144, 172-173
Petronzi, R. 172
phobia 7-8
physiological measures 35-38, 163
Pieronkiewicz, B. 122
Pizzie, R. G. 177
place value 116-117, 125-126, 129, 136
Pletzer, B. 38
Plummer, C. 113
Popham, W. 73
preconceptions 29, 100
prevalence 2, 19-23, 25-26, 28-29, 32, 35, 203
Primi, C. 51
problem-solving 10-11, 38, 48, 50, 58, 60, 64-70, 72, 114, 119, 123-124, 130-131, 149, 150, 154, 156-157, 161, 165
procedural knowledge 117, 127, 130
Processing Efficiency Theory 68, 70
Programme for International Student Assessment (PISA) 24-25, 30, 40-41, 108
psychodrama therapy 141, 170-171
Punaro, L. 16
punishment 5, 29, 79, 112, 120, 156

qualitative approaches 35, 52-54, 185
questionnaires 2, 21-23, 35-36, 39-55, 158-159

Ramirez, G. 69, 167-168, 170
Rapson, R. L. 104
ratio 127-128
Rattan, A. 115
reappraisal 141, 176-179, 184
Reeve, R. 16
Reeves-Kazelskis, C. 47
relaxation 3, 140, 142, 160-166, 176, 184
rewards 79-80, 120-121, 156-158, 170
Richardson, F. C. 10, 20, 45
Ridley, K. S. 20
Rodkin, P. C. 104
Rodriguez, C. 26
Rossi, S. 64
rote memorisation 117, 125, 130-131

Sachs, J. 46
Salazar, L. R. 163
Sammallahti, E. 144, 166
Samuel, T. S. 161, 179
Sari, M. H. 97, 195
Sarkar, A. 74
Scale for Early Maths Anxiety 51
Schalkwyk, G. 172
Segumpan, L. L. B. 150
self-awareness 157-160, 184, 201-203
self-belief 1-2, 12-13, 58, 63-64, 73-78, 81-82, 86-89, 97-98, 107-108, 110, 112-114, 116, 123-125, 137, 152, 173-176, 178-179, 185, 191-192, 195, 198, 201-202
self-efficacy 41, 71, 77, 180
shame 42, 153, 178, 203
Sheffield, D. 37, 47-48, 70, 141, 144, 165
Sherman, J. A. 46
Shin, Y. J. 105
Short, D. 25
siblings 3, 12, 27, 54, 109-111, 136
Silver, A. M. 102
Skemp, R. R. 82
social norms 103-107
socio-economic status 21, 23, 25-26, 32, 99, 104, 111, 143, 201
Soll, B. 173
Soni, A. 95
specific mathematical concepts 3, 12, 59-60, 116, 125-137, 146
speed 83, 111, 118, 121, 124-125, 131, 146, 157

Spinath, B. 80
standardised tests 59-60, 64, 145, 194
state anxiety 10, 12, 14, 31, 36, 163, 169
statistics 20, 30, 32, 48, 105, 132-133, 145, 171
stereotype threat 12, 31-32, 197
Stoehr, K. J. 14, 59
stories 3, 99, 141-142, 153, 171-172, 184, 190
stress 8, 28, 36-37, 55, 83, 85, 94-95, 98, 104, 114, 122, 124, 153, 157, 160-161, 166, 170, 172, 176-177, 181, 189, 192, 194, 201-202
subtraction 128-131, 147
Suérez-Pellicioni, M. 156
Suinn, R. M. 10, 20, 45, 49
Sunter, S. R. 108
Supekar, K. 146
systematic desensitisation 3, 141-142, 164-166, 176, 184
Szczygieł, M. 122
Szucs, D. 97

Tan, D. A. 150
Taylor, S. 42, 49
teachers 1-7, 9-10, 12-14, 19, 21-23, 27, 29, 31, 43-46, 48, 52-54, 59, 64, 67, 69, 71-72, 74, 76, 78, 80, 82-86, 88, 93, 95, 97-98, 100-101, 103-104, 106-108, 111-123, 125-126, 135-136, 140-141, 143-144, 146-160, 162, 164-166, 168, 170, 174, 176, 178-181, 183-185, 189-198, 200-201, 203-204
teachers' Maths Anxiety 190-198
test anxiety 15-17, 43, 45, 50, 64, 122, 153
Thilmany, J. 20
Thomas, G. 51
timed situations 3, 16, 60, 93, 122-125, 136, 162
Tobias, S. 11
Toffalini, E. 97
trait anxiety 10, 12, 14, 36
Trujillo, K. 11
Tunku Ahmad, T. B. 21

Tuohilampi, L. 73
twins 27-28, 110-111

Utami, S. W. 175
Uusimaki, L. 193

Vahedi, S. 150
Vallerand, R. J. 40
van der Vleuten, M. 109
Verleger, M. A. 149
visual aids 131, 156, 180, 183
Vokatis, B. 173
Vukovic, R. K. 70, 134

Walter, H. 167
Wang, Z. 27, 83, 110
Warner, J. 161, 179
Weesie, J. 109
WEIRD countries 23-24, 43
Weissbrod, C. 11
Westenberg, P. M. 108
Widjajanti, D. B. 75
Wigfield, A. 46, 77
Willingham, D. T. 10, 123
Willis, J. 148
Winston, E. H. 45
working memory 2, 11, 17, 36, 38, 58, 62, 64-71, 83, 88-89, 122, 124-125, 129-130, 132, 136-137, 154, 156, 167, 201
written methods 114, 129-131, 141, 156
Wu, S. S. 38, 51

Xenidou-Dervou, I. 64

Yanuarto, W. N. 12, 117
Young, C. B. 38

Zakaria, E. 79
Zhang, M. 25